The Right Way to Keep Chickens

The Right Way to Keep Chickens

by

Virginia Shirt

RIGHT WAY
plus

Typeset in 11 pt Legacy Serif Book by Letterpart Ltd., Reigate, Surrey.

Printed and bound in Great Britain by Mackays of Chatham.

The *Right Way Plus* series is published by Elliot Right Way Books, Brighton Road, Lower Kingswood, Tadworth, Surrey, KT20 6TD, U.K. For information about our company and the other books we publish, visit our website at www.right-way.co.uk

Chicken illustration on the cover by
dizzy@houseinthewood.f9.co.uk

Dedication

This book, my first publication, is dedicated to My Beautiful Mum. You always believed in me and encourage me in all that I do. Mum I love you so much and will love you forever and one more day.

Contents

1.   Preparing for Your Chickens   11

Housing, 11; Drinkers, 13; Feeders, 15; A Place to Lay Her Eggs, 17; Perches, 19; Bedding Materials, 21; Dust Boxes, 23; Mr Fox, 24.

2.   Keeping the Chickens Happy   27

Free-Range Chickens, 29; The House, 30; Clipping Wings, 32; Back Garden Chickens, 34; The Fold Method, 36; The Eglu, 37; Deep Litter, 39; Straw-Covered Yard, 41.

3.   Purchasing Your Chickens   43

Breeds of Chicken, 43; Bantams, 47; What Age to Purchase, 49; How Many Chickens? 50; Where to Buy Chickens, 50; Transporting, 51; Handling Chickens, 52.

4.   Feeding Your Chickens   55

Digestion in Chickens, 55; Feeding Extras, 60; Mixing Your Own Rations, 62; Vital Food Contents, 63; Vitamins, 64; Supplements, 65; Water, 66; Calcium and Grit, 67; Control of Vermin, 68; A Little Bit About Rats, 69; Poison, 69; Rat Traps, 72; To Summarise on Feeding, 72.

5.   Care and Use of Your Eggs   73

Egg Collection, 73; Storing Eggs, 75; Freezing

Eggs, 77; Preserving Eggs, 78; Pickling Eggs, 79;
Good or Bad Eggs, 80; Using Your Eggs, 82;
Recipes, 84; Eggs for Beauty Treatments, 87.

6.   Extra Joys of Chickens   89

Where do Eggs Come From? 90; How to Spot the
Broody Hen, 92; Breeding Replacement Chicks, 94;
In-house Hatching, 97; Using a Brooder Coop, 100;
Handling Chicks, 101; Rearing Chicks, 101; Feeding
Chicks, 103; Vaccinations, 104; Chicks: a Summary,
105.

7.   Health and Disease in the Chicken   106

Cleanliness, 106; The Five Freedoms, 108; Signs of
Good Health, 109; Signs of Poor Health, 109;
Vaccinations, 122.

Index   125

Illustrations

1. Make your own chicken coop   12

2. Types of drinker   14

3. Types of feeder   16

4. Ideas for home-made nest boxes   18

5. Battery hens   28

6. Free-range chickens   31

7. Clipping the wings   33

8. A fold unit, and the 'Eglu'   38

9. The Warren   45

10. The Orpington   47

11. The Bantam   48

12. Picking up a chicken   52

13. Holding a chicken   53

14. Chicken's digestive system   56

15. The position of the crop   57

16. Dealing with rats   71

17. Rotating the eggs   76

18. Candling eggs for freshness   82

19. Reproductive organs of the female chicken   91

20. A broody hen   93

21. Healthy chicken   110

22. Ill chicken   111

Chapter 1:

Preparing for Your Chickens

It is important to be well prepared for the arrival of your chickens. Once you have decided to make the commitment, you need to have everything ready. You may be lucky enough to know where to purchase your feeding and watering utensils; if not, most feed merchants will happily advise. You will find that most farming suppliers dabble in a few poultry sundries. Pet shops and equestrian outlets may also supply the feed and sundries.

Housing

Depending on the number of chickens you are planning on keeping, you will need to have a suitable-sized house or coop for them. There are many ready-made chicken houses on the market today. Some of them are really lovely but they are also very expensive for what they are. If you are handy with wood and a saw, then it is possible to make your own, and Fig. 1 (overleaf) shows one possible design.

I have used empty buildings and converted them into chicken housing. The conversions have been simple: nest boxes, perches, and a light and there you have it. Chickens do not require anything posh or fancy, making it easy to do on a budget.

If you do have a spare shed or similar, the process is easy. You can adapt almost any outbuilding or shed. Some farmers have even roughly converted the back of an old van or even a caravan. These are suitable if your chickens are

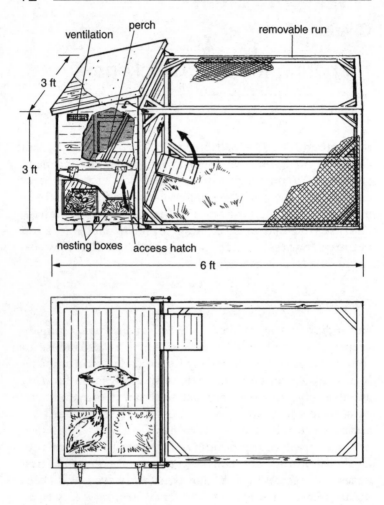

Fig. 1. Make your own chicken coop

This can be made any size; the one shown here, about 6 ft by 3 ft (2m by 1m), will suit about 8 chickens.

free-range as they will be out in the day anyway. They would merely be entering the accommodation in order to roost securely and lay eggs. Obviously with a caravan or similar you must not be tempted to overcrowd. This method will cater only for a small flock of eight/ten chickens.

Drinkers

Drinkers and feeders come in various sizes with drinkers holding one to four gallons. Again this is dependent on how many birds you would keep. The water needs to be fresh every day; it is no good having a four-gallon drinker for half a dozen birds. It is better to have a smaller drinker that is refreshed daily.

You can serve water to the chickens in an ordinary bowl or trough but the water will become soiled and contaminated. The chickens tend to do droppings in the water if it is not covered. Bell drinkers designed for poultry enable the birds to drink the water without fouling in it. The water containers can be obtained in plastic or galvanized steel. The plastic ones are adequate and cost a lot less than the galvanized ones. It depends on your budget; galvanized will outlive the plastic and therefore you may find it worth the investment. I have used the bell drinkers with success but I do find them rather hard to fill. These plastic bell drinkers need to be filled upside down (this is difficult in itself as you have somehow to balance the drinker and fill with a bucket at the same time). Once full, making it heavy, you then need to put the top on and tip it up the right way without dropping the whole lot! There is a bit of a knack to this particular job.

Automatic bell drinkers are ideal for large numbers of chickens. These save a lot of labour for the keeper. One automatic bell drinker will serve up to one hundred birds. The bell drinker is designed in such a way that the birds cannot sit on it and it is suspended from the roof by a chain.

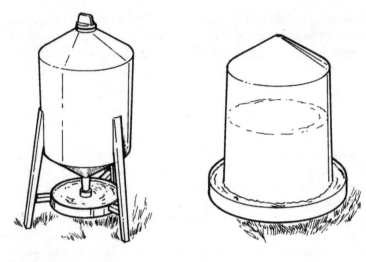

Fig. 2. Types of drinker

A tripod drinker on the left, and a bell drinker, of the non-automatic variety, on the right.

The water pipe comes down parallel to the chain and inserts into the bell which contains a ball valve. This method will provide a constant fresh flow of water.

A tripod drinker proves popular with poultry. This is a barrel-shaped water container standing in a tripod. It holds a large amount of water; the water feeds into a small bowl. When the bowl is full, the water stops flowing due to the vacuum; as the water is drunk, more is let through into the bowl. I find the barrel easier to fill than the bell drinker. It fills even easier if you use a funnel. This is my favourite drinker as it is easy to manage.

A nipple drinker is also labour-saving for the keeper. This is a very simple piece of apparatus. Fitted to a plastic water pipe, the pipe is drilled and the threaded nipple screwed in. When the bird presses the nipple, water is released. When the bird has drunk sufficiently, the water pressure shuts the nipple off. These drinkers can tend to have a leakage problem from time to time. In light of this problem make sure you site them away from food and roosting areas.

Feeders

Feeders are also designed so that the chickens can eat the food without contaminating it. Tube feeders are available and provide for all the needs of a chicken feeder. These keep a small amount of food in the base. As the chickens feed, the tube allows more food to drop into the base. Tube feeders reduce waste and contamination greatly.

A feeder needs to be placed at such a height that vermin cannot get at it but the chickens can! It also needs to be at such a height that the chickens do not scatter and waste the food. Place it at chicken neck height so the birds can eat comfortably. Feeders can usually be hung from the ceiling of your house or coop. If you cannot hang them up, they can be placed on concrete slabs or similar. Do not fill the feeders up to the top unless they have a lid on; fill them to about half way. If you fill open ones to the top, the chickens may jump on and feed from this point. This would lead to waste and contamination.

When deciding the feeders to purchase we should consider how much feeding space needs to be available to each bird. Although the birds will not all feed at once, we need to be sure they all can if they wish. To avoid bullying and chickens being kept away from the feeders, it is suggested that about one inch (2.5cm) per bird is made available on a round feeder.

If you decide to use a trough, then four inches (10cm) per bird is advised. Birds around a trough are on a straight line and therefore need to have their whole body size catered for. On a round feeder we only really have to consider space for their heads as they are on an angle. Troughs are available for chickens but you must make sure they have a cover on to avoid any soiling of the food. If a trough is twelve inches (30cm) long, then you have twenty-four inches (60cm) available for feeding the birds provided all sides are away from the wall. If you place the trough against a wall, then the trough will serve only three birds as opposed to six when placed away from the wall.

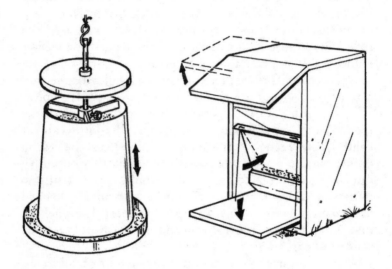

Fig. 3. Types of feeder

A tube feeder on the left, and a treadle feeder on the right.

If you do feed a lot of vegetables, then a small rack or trough should be used. These will be available from your supplier. It is important not to throw food onto the floor as this can cause dirt and bacteria to contaminate the food. You could use a heavy pan or bowl to place fresh vegetables in. I have a Mexican hat, which is a large, old-type pig-feeding bowl. It is very heavy iron but ideal for throwing vegetables into. All the birds can fit round to get at the food. If you know of any pig keepers selling sundries, it is an ideal piece of equipment to have!

Food and water should, if possible, be served indoors. If the food is served outside, wild birds and animals could eat it. Not only would this be uneconomical but it could also place the chickens at risk of disease. However, if feeding outside, then the treadle feeder is an ideal choice. The food is closed away inside a container protected from other birds or vermin. This feeder requires the chicken to stand on a treadle to receive food. The food is also kept dry and clean, allowing it to be fed in large quantities without waste.

A Place to Lay Her Eggs

It is very important to have a suitable set-up for the chicken to lay her eggs. Nest boxes can be purchased but again can be expensive. If you have any old cupboards or wooden boxes, they can be put to use for egg laying. I have also noticed that chickens love to lay eggs in an overturned bucket or dustbin! They even like to lay behind old leaning doors or upturned rabbit hutches, etc. You can simply improvise and convert items that provide a dark, cosy and warm environment. You don't have to spend lots of money, just use your imagination. If you cannot find anything suitable around the home, you may decide to make your own (as shown in Fig. 4). A bit of wood and a few nails and you can knock up the most delightful nest boxes!

The nest boxes need to be homely and inviting to the chicken. I have heard it argued that the nest boxes should not be too comfortable as they may encourage broodiness. However, I have not come across any real evidence for this. You need the nest box to be clean and well bedded to prevent too much soiling of the eggs. I also feel that, if the chicken is working daily to give me an egg, then she should pass the said egg in a little comfort!

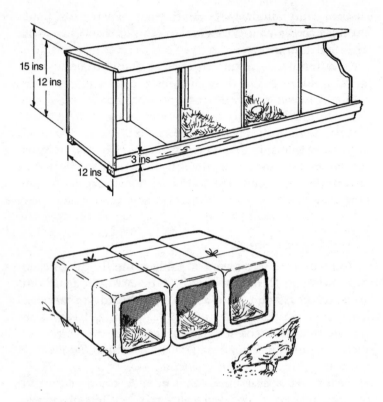

Fig. 4. Ideas for home-made nest boxes

Make your own nest box (above), or improvise (below).

Free-range chickens will find the most unusual places to lay an egg. Bowl-shaped items that they can fit snugly into are ideal. They love bales of hay and straw to nest in. If you have a hay barn or a hay storage area, you will find that this is a favourite spot for your chickens to lay. I have even found eggs in a redundant cat basket!

I recently purchased some chickens that were raised in a milk churn that was on its side. Such an object provides a chicken with all her brooding needs. There is a small entrance to the churn; this makes her feel safe and secure. The small entrance will also keep out the cold. The inside of the churn provides a snug area with a draught-free environment, an ideal spot to raise her young.

If you make sure your nest boxes are comfortable, safe and secure, this will encourage the hen to lay in them. If the hen lays in the correct place, you will be able to collect each day without having to hunt around. Place the nest boxes in the darker areas of the chicken house, as hens do like a dim area to lay. With the nest boxes in the darker area of the house and the food and water in the well lit areas, the hens will be discouraged from laying eggs on the floor. Raise the nest boxes off the floor to deter the rats from entering. Rats do love a freshly laid egg!

You will need to provide at least one nest box for every two to three chickens. If you do not provide enough laying areas, the hens will squabble over the nesting area and eggs may get broken. Remember that we should avoid broken eggs at all cost. Once the chicken tastes what is inside the shell it is difficult to discourage her from breaking the eggs in future.

Each nest box should be approximately one foot (a third of a metre) square. Any smaller and she is cramped, too much bigger and she does not feel snug. Also it is better to have a size that fits only one chicken in at a time. You do not want to encourage two or three at a time to squeeze in it. Too many laying in one nest box at the same time can cause the eggs to get broken or cracked.

Perches

Chickens need to perch at night and so we must provide the perches for them. You need enough perch space for each bird, this will be just under a foot (30cm) for each. I have noticed that some of the chickens really like to snuggle up close to another. Some like lots of space between each of them. I suppose they are all individuals, just as we are.

Perches need to be removable so that they can be cleaned once a week to avoid a build-up of droppings. If the perches are kept clean, this will also keep the chickens' feet in good health.

Perches need to be high enough from the floor to keep the chickens out of the draught. A good height would be around a foot (30cm) from the ground. This is an ideal height for them to leap onto from the floor, and also makes getting down in the morning easy. A maximum height for the perches would be two foot (60cm). If the chicken had to jump down from a perch higher than this, she could do damage to her internal organs. She could also be susceptible to a condition called bumble foot, which is caused by the *Staphylococcus* bacteria entering the blood stream. The bruising caused by jumping from a perch will predispose the chicken to bumble foot. Landing heavily on the feet could also cause a small entry wound for the bacteria to make their way in (see Chapter 7, Health and Disease in the Chicken).

A builder's lathe or a piece of wood one inch (2.5cm) square would be suitable for a perch. It is just the right size for the chicken to grip with her toes. The stake must be sanded down so that it is very smooth. A smooth finish will be kinder to the chicken's feet; also mites will not be able to hide so well in it.

If you have an old set of ladders, these can be used as perches. Ladders are generally a smooth finish. They are an ideal size for the feet of the chickens to hold onto.

Chickens have a special locking system in their legs which enables them to sleep and yet remain perched. The chicken will not fall off even when she is resting; when she grips onto the perch, the muscles and ligaments in her legs lock, thus preventing a fall.

The perches also enable the birds to roost without being

in the dirt. Should a chicken sleep on the floor she might become sore from the ammonia, etc., from the droppings. For health reasons this would not be advisable. Sufficient perching area will also prevent the chickens from sleeping in the nesting boxes.

Bedding Materials

Shavings

I have found shavings to be the most efficient of all the bedding materials I have tried. They are warm and soak up any moisture caused by the chicken. The chickens also like to scratch around in the shavings and take a daily dust bath. Although this is not really of great importance to the chicken, shavings do look pleasing to the eye when freshly laid.

Shavings are also economical and a big bale can be obtained from the feed merchant. If you have a local wood yard, you may find they will supply them cheaply. Wood shavings are ideal with a deep litter system as you can simply add to the bed as and when the need arises. The only time I would not advise shavings is when raising young stock. Initially for the first few weeks of life, the young chicks may mistakenly ingest the shavings, thinking they are food. This can lead to serious health problems and even death. Also check out the weight of the bales. I frequently find a big price difference when shopping around for shavings. I have often found them at a low price only to discover that the cheaper ones are not so tightly packed. Also, do not be tempted to buy shavings that are marketed for horses. Again the bales are very light in comparison to other bales.

Wood Chips

I have found wood chips to be a little more expensive than shavings. Not only do they cost more but also seem to cover less of the floor than the shavings. I have also found them to soil at a faster rate than shavings. The only advantage wood chips have over shavings is that they are safer for young stock. As the wood chips are larger and more coarse, the chicks will be less likely to peck at them thinking they are food!

Shredded Paper

Paper is not recommended for use as bedding material. Initially it looks pleasant and gives a nice warm feel to it. But it will soil quickly and, when wet, will become solidified and difficult to manage. However, on a good note, paper can be cheap if you shred your own. You could recycle your own paper and that of friends, etc. If you can manage the extra work involved in the use of paper, then it is economical when recycling your own paper supply. For all its problems, paper is a good bedding when raising young chicks. Although it does soil quickly it is cheap enough, when you are recycling your own, to change every day. It is also very insulating and can help the young stock to maintain a warm environment.

Hay

Hay can be used in nest boxes but is of little use as a bedding material. It has poor drainage properties and would not provide good scratching and bathing material for the chickens. I find that it is the best bedding material for the nest box as it provides a cosiness that moulds to the shape of the chicken. However, many chicken keepers prefer not to use hay in the chicken house as it is said to bring in the mites and fleas.

Straw

Straw does have better drainage properties than hay, particularly wheat straw. It is a good material for the chickens to scratch around in and can be pleasant in the nest boxes. It will also make a good litter for the floor of the chicken house. To save you the job of turning it over with a fork, throw some corn over it. This will encourage the chickens to work over the straw for you. Straw will work better if it is chopped into smaller pieces. Straw is a cheap bedding material and is readily available from local farms. For the same reasons as for hay, many chicken keepers will keep the hen house free from straw, as they believe it encourages mites. With straw being tube-like in its form, it also provides good shelter and a breeding ground for the mites.

Don't be tempted to buy any of your bedding material from pet shops. Pet shops can provide all the beddings that are required but they sell small quantities for a large amount of money. Purchase in bulk, either from farms or feed merchants, etc.

Dust Boxes

Providing some form of dust box or bath will supply the chickens with endless fun! If your chickens are free-range, they will probably create their own dust bathing area. However, if your chickens are contained in any way it is kind to provide the bathing materials. Watching a group of chickens enjoying a dust bath is very amusing; they get so much pleasure out of it. You can make it out of a large, flat litter tray that you would use for a cat. Any similar shallow container will do, so long as it can fit a few chickens in. Fill the container with white sand and away they go. Remember also that this is an ideal way to administer flea treatments just by adding the flea powder to the dust box. This is explained on page 119.

Place the dust bath in a sunny area; chickens do so love the sunshine. Don't leave it there when it rains. You need the sand to remain dry for the bathing. Although free-range chickens will usually find a bathing area themselves, you may find the need to provide one. Again this must be placed in a sunny area. If you have lots of chickens, then three or four cat litter trays, or similar, should be provided. Make sure that the dust bath is kept clean and dry; refresh frequently.

Mr Fox

Never forget our number one enemy! He is our worst nightmare and must be kept away from our chickens at all cost. He is indiscriminate and seems to kill just for the fun of it!

Although many people believe he strikes at dusk and dawn, he also often pops by on an afternoon.

He is like a bailiff; he has no particular calling time. Many of my fellow keepers have had all their chickens killed during one visit. I have been lucky and usually lost only one or two at a time. The worst incident I had was when I heard my chickens' hysterical calls. When I ran to them I must have disturbed the fox. I found four bodies then, which was bad enough.

Fox Deterrents

If you own dogs, they can be a deterrent to the fox. When walking your dogs, take them around the perimeter of your land or garden so they leave a scent. As unpleasant as it sounds, collect the faeces left by your dog and place them around the perimeter or entrances of your chicken area. Place at regular intervals or near entrances to your garden, property, etc. It may not be practical to use this method if you have neighbours close by!

Another deterrent is human male urine, although I am not sure if it would be easy to deposit this round your perimeter without being arrested! It depends on how desperate you are to keep the fox away. I haven't managed to persuade my husband to oblige me in this area yet. However, many years ago I did hear of a chicken keeper who would never urinate in the toilet. He always did it in a bottle and saved it for this very reason!

Human hair is said to keep the fox away, as they do not like the scent of us. A visit to your local hairdresser should provide you with a steady supply. Stuff the hair into pairs of old tights and tie it to your fence at intervals of about a foot (30cm) apart. This will depend on how much hair you manage to get hold of.

Any form of fencing you put up around your chickens must be dug in one foot (30cm) below ground level. Foxes will try to dig up the ground to get into the chicken run. Chicken wire can be obtained from most poultry suppliers or DIY stores and comes in fifty-metre rolls. It is usually three or six feet (90cm or 180cm) high. I would advise six feet (180cm) high to be the safest option. You can also purchase electric chicken mesh; this comes in a very bright orange or green. This would be an ideal form of fencing but you must always make sure that the fence is live (making sure electric volts are going through). Animals will soon detect when an electric fence no longer gives off a current.

If you have any rabbit hutches, etc., for sick bays or chick rearing, they must be locked securely at night. Rabbit hutches are brilliant for the above purposes but do not keep the fox out. Bring them into the garage, barns, stables or even the porch of the house. Do remember that the fox may come during the day so keep them close to the house all the time, and be vigilant.

If you use an electric fence, the wire needs to be 3–4 inches (8–10cm) above the chicken mesh and kept tight. Tape or wire is acceptable. The fox can put his head through but will get a shock.

Chickens do not seem to require a great deal in exchange for their lovely eggs. Once the initial preparation has taken place they require little attention compared with many other animals. The initial setting up will require a substantial financial outlay, but if you look after your equipment, etc., most of it should last for many years. If you do find the outlay expensive, think about advertising in the local paper for second-hand equipment. If you renovate, clean and sterilize second hand goods well, they should be as good as any new items.

Chapter 2:

Keeping the Chickens Happy

We need to give proper consideration to the various methods of housing the chickens. For the chicken enthusiast there is nothing to beat watching a happy, healthy chicken, running around and living life to the full! Sadly, not all chickens are able to do this. Some are kept in such cramped conditions they cannot even exhibit normal behaviour.

Chickens need more in their life than eating and laying eggs. They need to socialise and form their little groups. They need to bask in the sun and stretch out playing dead. They need to roll around in a dust bath. They even need to fight and argue, sorting out the pecking order. They need to run, jump, flap and shake.

Unfortunately, the life of a battery hen allows for none of the above (see Fig. 5). It is simply to eat and lay eggs. It is heartbreaking to think of the chicken unable to live normally. Serious mental health issues develop in the form of cannibalism and feather pecking. The hens are also depleted of calcium, leading to lameness and even broken bones. Many battery hen rescue organisations actually have to train the rescued chickens to walk!

Battery hens are cramped into cages in threes and fours, which is unnatural and unfair. They will often be de-beaked in order to prevent injuries from stress-related aggression. The de-beaking procedure is painful in the first instance. Distressingly, the pain does not subside but leaves damaged nerves from the barbaric practice. The purchase of free-range eggs is on the increase; hopefully one day the battery hen will become obsolete.

Fig. 5. Battery hens

What a miserable existence!

The majority of people reading this book will keep chickens for pleasure and, of course, eggs. We will, therefore, explore the less cruel methods of housing and keeping chickens. There is a method to suit most budgets, lifestyle and property.

Poultry housing does not need to be costly or fancy. The most important requirements that the chicken accommodation should provide are:

- Shelter from the elements
- Protection from predators
- Nesting boxes
- A place to roost
- Freedom to exhibit normal behaviour.

Free-Range Chickens

Keeping chickens on the free-ranging system is one of the most idyllic and popular options. True free-range chickens are allowed to roam freely over many acres with no fencing of any kind. Chickens do not tend to take advantage of that much space, they prefer to stay around the housing area. The free-range chicken should be fed and watered outside and only come in to the chicken house during extreme weather conditions, and for roosting and laying eggs.

Free-range eggs have been shown to contain two/three times more beta-carotene than factory eggs; four times more omega 3 fatty acid; twice as much vitamin A. Cholesterol in the free-range egg is tested to be half the factory egg content (*Mother Earth News* magazine).

The true free-range chicken derives most of her food from the land. This would include insects, slugs and worms. Of course, the grassland and vegetation itself makes up a large slice of the diet. The health of the chicken is much better living freely. Ventilation is perfect and disease outbreaks reduced with less confinement. Labour is reduced, as is the feed bill. With completely free-ranging chickens you will not receive as many eggs; however, if you fed layers pellets alongside the natural diet, you could restore the yield of eggs.

Behaviour problems are eliminated by the freedom to roam. Cannibalism is wiped out, as is feather pecking. Aggression is reduced, as the chickens keep out of each other's space. Stress is the cause of most behaviour problems; free-range chickens suffer little if any stress.

However, there is a major disadvantage to the truly free-range set-up. That is the loss of chickens to predators. You may find that foxes and badgers will pay visits to your flock. Once such predators realise they have a regular source of food, they will return with increasing frequency. So the question of whether you can countenance a fully free-range arrangement depends on two factors: firstly the number of predators in your area, and secondly your own attitude to the loss and damage that predators will cause. If losses are few, and your response to them is not unduly emotional, then maybe unlimited free-range is for you.

Chickens can still be described as free-range, even when fenced to some degree. In an ideal world, chickens would roam freely; we do not live in an ideal world. We owe it to our chickens to give as much protection as we possibly can. I feel that to lock chickens away for twenty-four hours a day, seven days a week is unkind. Is it not better to live a shorter life with freedom rather than a very long life confined inside four walls?

Therefore I support semi-free-range, keeping the chickens as free as possible yet with as much protection as we can give them. You may still lose some of your stock on odd occasions but sadly that is almost inevitable. I keep them in a big stable at night and they are out all day. I confine them to a degree, finding that this reduces my losses. They are able to roam about the place but within outer boundaries (of walls, wire mesh fences, etc.). I must say they always seem to be rather content. I found it too traumatic losing chickens on a near regular basis. You just never know when the fox will call.

The House

Depending on how many birds you require, the house must be of a suitable size. The largest poultry house commonly available will house up to 200 birds but a smaller version is shown in Fig. 6. The house will be on skids with fitments where a chain can be threaded. The chain can then be attached to a tractor and the house then pulled easily to its weekly destination. Do not move the house too far each time; the chickens must not lose sight of it or they may get confused as to where to go at night. Just move it far enough to freshen the ground and rest the old area of land.

Fig. 6. Free-range chickens

The house needs to be moved to new ground from time to time.

The nest boxes are designed to allow the eggs to be collected from outside the house, which makes egg collection much easier. Two or three pop holes are advisable to allow entry to all chickens. With just one pop hole, low-ranking chickens may be barred from going in at night by the higher-ranking chickens. Having extra pop holes usually prevents this problem.

Pop holes should be away from the prevailing wind, to inhibit draughts and rain coming in through them. The pop hole needs to measure a foot (30cm) square. It only needs to be big enough for a chicken to go in and out. Too big and it will allow not only the cold in, but, more importantly, predators. I know of someone who had an oversized pop hole which allowed a badger to enter. Sadly nobody knew the badger was inside and the pop hole was shut for the night. You can imagine the devastation! The pop hole should be opened at dawn by pulling up the drawstring and hooking it up securely. Unhook the door and drop the pop hole down when all the chickens are inside at night.

Not everyone can be available at dawn and dusk. There may be work or social commitments that make such a tie undesirable. There is an alternative, which can make life a little easier. You can purchase an electronic doorkeeper, which can be used to open and shut your pop hole during any absence. The device can be set on a timer, which you can alter in keeping with the changing times of dawn and dusk. You simply attach the unit to the pop hole and it runs off batteries. While I think this is a good option to use if you have no alternative, it is by no means ideal. Often when dusk arrives, for some strange reason chickens can occasionally decide to roost outside the house. It would be sensible to check round with a torch when you do arrive home just to gather up any absentees. Also take a look inside the house just to make sure all is well and that no predators have sneaked in for the night.

Chickens do not like extreme weather conditions, and so shelter must be available for free-range birds. I often notice that in really bad weather the chickens do not return to the house. Instead they will just hunch up and sit it out, looking very unhappy. If you have walls, bushes, trees, etc., this provides some natural shelter. Chickens soak up the rain like a sponge and are not waterproof!

Clipping Wings

If you keep chickens free-range, with no fencing at all, then you should not clip their wings. Being able to fly a little will help them to get away quickly from the fox. However, with semi-free-range you will probably have a fence to keep them secure. In that case, the wings will need clipping, so that the birds do not fly over your fences and boundaries. It is a simple enough task and causes them no pain. You clip only one wing on each bird and that will take away their ability to become airborne (not that chickens are good at flying but they can flap well enough to clear a fence).

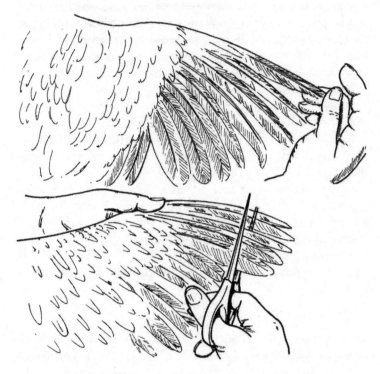

Fig. 7. Clipping the wings

It's like cutting your toenails.

Gently taking hold of the chicken, span out her wing as shown in Fig. 7, then cut with scissors (preferably with rounded ends) about three inches (8cm) off as the lower picture shows. You will notice that the wings being clipped are secondary feathers. These are not required to keep the bird warm like the primary feathers. The secondary feather shafts (the long bit that runs through the middle of the feather) are a light colour. The darker colour indicates blood supply and you must, therefore, avoid cutting in these areas. If you are unfortunate enough to draw blood, the hen must not be put back in the flock until bleeding has ceased. If you do place her in the flock, she could be attacked and even killed. I would advise dabbing on some veterinary wound powder or similar. This will help dry the area and reduce inflammation.

If you liken cutting feathers to cutting your toenails, you will understand the principle. When you cut your toenails, you only cut the transparent bit. If you cut too low into the darker (pink in our case) colour, you cause pain and draw blood!

Clipping one wing only, to cause loss of balance, results in an inability to leave the ground to any great height. This job will need doing about once a year, after each moult. Young stock will moult more often during their first year but clipping should not really be carried out on chickens less than 6/8 months old, as the feathers are still growing.

Back Garden Chickens

You can keep the chickens semi-free-range in your back garden if you live in a town. In a small garden you could consider half a dozen chickens with a small chicken house. Enclose your garden with chicken mesh, and/or electric fencing 5/6 inches (13/15cm) from the ground. Such a method should keep the chickens in and the predators out. Farm and poultry supply shops sell some very good electric fencing specially designed for chicken enclosures.

Obtain a small chicken house (about the size of a Wendy house). This should then be kitted out with nest boxes and perches. You could place the food and water in or outside the house. Outside would be preferable to avoid taking up valuable living space. Do keep the food and water under cover; only feed the required rations to avoid waste and prevent vermin, etc. A treadle feeder would be ideal.

Never consider free-range or semi-free-range chickens to be safe from the fox. You must always remain vigilant even if you do have the required deterrents. Never ever leave chickens out all night. They must be locked up safely and securely, with all pop holes closed!

If you work during the day, you could limit the time your chickens have outside. You could let them out for a couple of hours when you get in from work. It is no good thinking of letting them go out for an hour before work, or during your lunch hour. One thing I have learnt with my flock is that they will not go back in until they are ready, no matter how long I chase them for. Chickens are also very perceptive and know when you are in a hurry. If you are trying to get them away quickly, they will sense this and be very awkward.

You may be able to get them in using scratch corn. Train them to respond to the rattling sound of the corn in a bucket. When you want them to go in, rattle the bucket. With luck, they may respond and enter the chicken house as you scatter the corn.

You may find it easier to let them out after work. They could enjoy the last hour or so of daylight (depending on the time of year). At dusk they will be ready just to be herded in. Letting them take their freedom at this time of day relieves you of too much pressure. You do need to bear in mind that dusk is also a favourite time for predators to pounce. Be especially vigilant, as you need to be at all times.

If you do have a dog, and your garden is dog proof, letting him out with the chickens can be a good idea, so long as your dog does not worry or upset them. Dogs will always discourage predators. Observe your dog each time you let him out. He may give signals that predators have visited. My dogs will often indicate that foxes have been around. They sniff anxiously around areas where I happen to know a fox has tried to get in. They urinate and defecate around the area that has been scented by the intruder. This behaviour is the dogs marking over the foxes' scent. It should signal to the predator that he is on claimed territory and should beware.

If I have evidence of the fox being present, I often change the routine of my chickens. If I am going out for the day, I will keep them in during this time. If you leave yourself in a position of dashing home before dark, it puts you under a lot of pressure. Leave the chickens in, making sure they have plenty of food and water to keep them occupied.

With a small garden, you should confine the chickens to small runs that are moved around the garden every few days. This prevents the land/garden becoming sour and contaminated and over-grazing is reduced.

If you are simply planning to have enough chickens to provide eggs for just you and the family, a tiny little ark or similar can be purchased. This is ideal for a small garden or back yard. The ark or cage is fully portable and houses a small number of chickens. There is enough room for laying eggs and roosting, with a place to exercise also. This system is ideal and very simple. You can even make your own little house for them. The chickens are let out in the morning and locked away securely at night. During the day they scratch around the garden, eating insects, etc. Three to four chickens should provide enough eggs for a small family.

The Fold Method

The fold method is even more ideal for the back-yard flock! The fold system comprises a small house with a larger run attached to it, as shown in Fig. 8. It is easy to manage and is effective against the fox and other predators. The chickens are completely enclosed inside the unit. The unit is usually on wheels or has carrying handles on the side. A very small fold unit can house as few as half a dozen hens (just right for the back-yard chicken keeper); while bigger units can be purchased to house up to thirty.

The unit is moved around as soon as the area beneath it is grazed off. Usually the unit is moved only as far as its own length. Moving regularly keeps fresh grazing available and prevents over-grazing and souring of the ground.

If you do only have a small area of land/garden, it is even more important that it be well maintained. Don't allow a build-up of droppings. When the unit is moved to a new area, the soiled area must be raked over. If the grass has been well grazed and is becoming thin, I would suggest re-seeding sparse areas. Re-seed with a mixture of grass; this could include clover, ryegrass, meadow grass, fescue and crested dog's tail. If you are really green-fingered, a variety of herbs adds to the diet of the chicken. I know from experience that chickens love the herb chocolate mint!

The area where the chickens graze benefits greatly from a yearly application of lime. This kills off parasites that could re-contaminate the chickens and allows the grass to utilise the nutrients more efficiently.

The Eglu

Although rather expensive, the eglu is an ideal addition to the garden of the city dweller. The eglu is an ingenious invention that is easy to clean and fox-proof also. The chickens have a warm and cosy area to roost and nest; they also have access to a small run to allow for exercise. The eglu comes in a choice of five different colours so there should be one to suit most tastes.

The eglu will house up to four large chickens or five bantams. It is easy to clean, making for a hygienic method of chicken husbandry. I personally think that they are ideal for city dwellers who simply want to eat their own home-produced eggs. It is made from a strong, durable material, which means that, although the outlay is expensive,

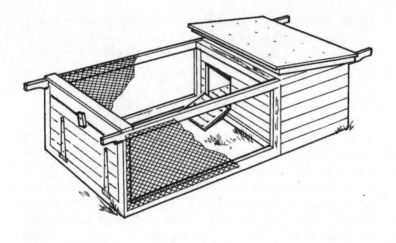

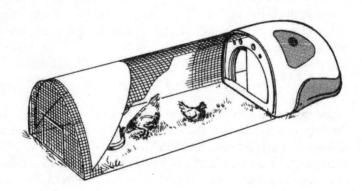

Fig. 8. A fold unit (above) and the 'Eglu'

the eglu will last virtually a lifetime if well looked after.

Deep Litter

The deep litter system is not in my opinion ideal. Sometimes, however, its use may be unavoidable. Not everyone has an abundance of land; bird flu being a possible threat means that many poultry keepers feel they have to adopt this method to avoid infection. Although with this system the birds are not free, it is more humane than the battery system. At least in a large building the birds can exhibit and perform the majority of their natural innate behaviour. I have found that when I keep my chickens in for several days, which is occasionally necessary, the chickens do adapt quickly.

With this system, the house you use must be as good as you can get it. There can be no half measures, as the birds will spend all their time inside this house. The house will need to be warm in the winter and cool in the summer. Ventilation must be adequate in order to keep the lungs of the birds working correctly. There must be a free flow of air passing through the building to keep the air fresh and clean.

With the deep litter system, loss from predators is non-existent. Egg production is high and you always know where to find the eggs. On the downside, disease spreads more quickly once a chicken becomes ill. The birds do not get the benefit of vitamin D from the sunshine. You need to make sure they are well supplemented with fresh greens, including grass and dandelions, etc. The design of the house should have a large meshed area that permits the rays of the sun to filter in to the chickens. My chicken house has a complete doorway made of mesh; this nicely lets in the morning sun. The sunrays not only provide vitamin D but also kill germs and bugs.

The chickens will require at least four square feet (60cm x 60cm) per bird to live a reasonably happy, fulfilling life.

Having the birds on less space than this leads to stress, resulting in feather pecking, cannibalism and aggression. Remember that, if a chicken does become ill, she must be removed immediately. If left in with the healthy chickens, she will soon be killed and eaten.

Starting the deep litter off is relatively expensive as several bales of shavings are required. The litter must be six or seven inches (15–18cm) in depth to begin with. It should be topped up regularly after this. Removal of all damp, soiled patches must be carried out daily. The deep litter provides insulation for the birds and keeps them warm. In warm weather they burrow into the bedding to cool themselves down.

You must keep the litter dry and well topped up to avoid damp and ammonia build-up in the air. The building needs to be cleaned out completely every six months or so. It certainly must be left for no longer than a year. Dig the surface weekly to keep the bedding fresh. Throw scratch corn for the chickens and they will turn the bed lightly for you.

You need to provide lighting for the birds if the windows do not let in sufficient light. Make sure that the lights are on a timer switch so that they come on for a fixed number of hours each day and go off at a certain time. Leaving lights on too long can lead to stress-related disorders for the birds and blood spots in the eggs.

Sixteen/seventeen hours would be the maximum light required. The hens do need the dark in order to rest sufficiently.

Each time the house is cleaned out, you have a very useful by-product. Chicken manure provides wonderful nutrition for your garden. It is considered by many to be better than horse and cow manure. Well-matured deep litter is also better than fresher chicken manure; it contains nearly three times as much nitrogen. It also contains two per cent phosphorus and two per cent potash.

Straw-Covered Yard

This system is a safe and secure combination of free-range and deep litter systems. The idea is to keep the chickens relatively free but with a degree of confinement. This system is chicken friendly and allows the chickens to lead happy and fulfilling lives. The chickens are kept in a building or shed with access to a large outdoor yard section. This is ideal if you do not have access to a garden or field, as the straw provides the chickens with a nice daytime surface to scratch about on. During bad weather the chickens can be kept inside, away from the elements.

Chickens hate to get wet. They are not waterproof like ducks. As soon as it rains, a chicken becomes wet through, and bedraggled, and likely to chill. During wind and rain, chickens should be kept in. You will notice that, if you do turn them out in extreme weather conditions, they stand cowering in a corner looking very miserable. If they have plenty of space indoors, it's wise to keep them in until a better day.

This system requires a large amount of straw to keep it pleasant and comfortable for the chickens. The straw must be kept topped up so that the chickens are on clean, dry flooring.

When considering the type of housing to use for your chickens, try to mimic what nature intended for them. I find that with any animal I have kept, and I have kept many different species, the closer to nature I have kept them, the less trouble I have had. To keep an animal in its natural environment will reduce the risk of stress and disease. When animals are taken from their natural habitat, this is when problems occur. Think about behavioural problems; many are the result of overcrowding and stress. Such problems would not occur in their natural habitat.

If we think about keeping them close to their natural state we must, therefore, reject the very idea of the battery hen! The battery system is purely commercial. It is to make money, with no thought at all for the needs of the chicken herself. Free-range is wonderful but opens up the possibility of predator attack. So the middle line is the straw-covered yard or semi-free-range. This provides a lifestyle close to the one that nature intended. If you keep maximum space requirements for the chickens, stress does not become a problem. Provided that the chickens have all their other needs met, this is a very good method to use.

Those with a garden or small amount of available land should use the fold method and limit numbers. Those with a larger amount of land available should use the semi-free-range method. Chickens have minds and needs, and the battery system should be completely banned. It is cruel, inhumane and unnecessary. We as consumers should only ever buy free-range eggs, as the chicken that produced the egg will be a happy one!

Chapter 3:

Purchasing Your Chickens

Breeds of Chicken

The first consideration is which breed you require. This depends on the job the chicken has to undertake for you. Some chickens are bred to provide eggs in abundance and little else. Others are selected to provide meat for the table. Then there is the chicken that fits both requirements.

This book is intended primarily for the keeper of chickens requiring eggs only. Therefore we shall concentrate on the egg layers. The breeds available are usually a hybrid variety; this is a cross of two different breeds or strains. They are specifically chosen for their ability to lay eggs throughout most of the year. Two birds crossed successfully will produce offspring with the best attributes of both parents. *Hybrid vigour* or *positive heterosis* is a term that describes the result of such a cross, which manifests in stronger positive characteristics of the offspring. The offspring will perform, hopefully, even better than their parents. A good cross combination can be repeated many times. Many of these hybrids produce chicks of different colour according to sex; these are referred to as *sex sal* links. This enables the females only to be reared on, resulting in cost efficiency. The males are sold at a day old to the pet industry as food for snakes, reptiles and birds of prey.

The first year will be the year of maximum egg production, but as time goes on the chicken will produce fewer eggs per year. This is why commercial keepers re-stock all their chickens at one or, at most, two years old. Don't worry too much about this seemingly short life span. I have often purchased older chickens and been rewarded with a plentiful supply of eggs. The commercial keeper is usually interested in only one thing: profit. Therefore, unless a maximum supply of eggs is given in return for the layers rations, then they have to go.

There are rescue organisations that can often supply you with some of these unfortunate chickens. If you look after them and feed them a good diet, then you will get plenty of eggs. It is particularly rewarding if you can rescue half a dozen battery hens. Although they often need rehabilitation, this can be a very satisfying way of keeping chickens. The ex-battery hens are very cheap. They need lots of love, rest and affection to recover from the ordeal of the hen house. These poor creatures normally spend one year in a battery cage and then they are slaughtered. Why not have a go and contact your local animal rescue branch or get details from the Internet? You may have a local chicken farm that will be re-stocking. Make enquiries at the farm and you may be able to save some chickens and give them a future. Even producers that do not keep chickens in battery cages will probably have a clear-out after one year. It is a scandal, really, when you consider the life that is left in such animals. They can go on producing for many years to come.

Don't get too serious about the amount of eggs produced. It's lovely to get a plentiful supply of eggs, but absolute maximum gain from the hens is not what it is all about. There are lots of attractive and appealing hens out there, so take a look around at the different varieties. I have kept a little trio of aged Silkies and have had very few eggs from them. When I came by them they were about to be slaughtered as they were no longer producing and were rather old. They just look so pretty strutting around the yard. When I do find a glossy little Silkie egg in the nest box it feels like a lottery win. However, you may not be able to afford or want to keep unproductive chickens; if so, you need to look at the higher-producing birds.

Fig. 9. The Warren

Usually a deep red colour, a lean laying machine.

Warrens (Fig. 9) were originally developed for the battery hen industry. They lay 250–300 eggs per year and carry virtually no meat on their body at all. All the energy from the food they eat goes into the production of the egg. Warrens are lovely chickens, usually a beautiful, deep red colour. These are the hens I started with and I found them to be very hardy and pleasant to work with. They are readily available for purchase at eighteen/twenty weeks old (point of lay) and seem to be the cheapest hens to come by. Other varieties tend to be much more expensive. If you are interested in breeding or showing your birds, you can look at the more expensive, fancier breeds.

Marans are an attractive addition to any flock and lay a lovely dark brown egg. The eggshell is very smooth and yet robust-looking. It has smaller pores than most eggs, which helps keep the egg more sterile. The egg white is denser; this helps it retain a good shape in the pan. The Maran does not produce as many eggs as the Warren, laying up to approximately 200 eggs annually.

The *Rhode Island Red* is a good layer, producing in excess of 200 eggs per year. This bird lays particularly well, considering it is bred for meat also. The eggs are a rich brown colour. The Rhode Island Red is also a hardy bird, making it easy to keep.

The *Black Rock*, which contains genes of the Rhode Island Red, is also a hardy bird. It has a very thick plumage, allowing it to enjoy most weather conditions. The Black Rock lays 250 or more eggs per year and these are brown eggs with a good quality shell. The Black Rock is believed to be less susceptible to red mite due to its quality feathers and plumage.

The *Orpington* (Fig. 10) originated from the town of Orpington. There are many different varieties of this chicken, my favourite being the *Buff*. Orpingtons are big, striking birds that seem to walk around very aware of their own beauty. They have exceptionally thick plumage, which adds to their impressive size. The Orpington hen lays around 200 light brown eggs in her first year. A nice, friendly bird to work with.

Just a small selection of breeds to provide food for thought. The breeds mentioned above are all good egg producers while being friendly and easy to manage. These birds are also easily obtainable from reputable breeders.

Fig. 10. The Orpington

A big striking bird, with exceptionally thick plumage.

Bantams

If you cannot spare the room for a small flock of chickens, then why not consider a trio of bantams? Bantams are basically small chickens that produce small eggs and require smaller accommodation. They can be a little more expensive to purchase than the ordinary chicken but you will reap your reward in the form of quality little eggs. Being smaller birds they will also eat less, of course. The only downside really is that they do not produce as many eggs as the commercial birds. However, if you look after them well and preserve your eggs for the famine periods I am sure you will not go too short.

Bantams are miniature chickens (Fig. 11) that are usually available as a small version of any of the larger chicken

Fig. 11. The Bantam

A Buff Rock Bantam (above); the lower picture shows how this bantam compares in size with the Warren chicken.

breeds. For the beginner and for children, the bantams are ideal. They also make nice, friendly pets that can be enjoyed by the whole family. The bantam can be housed in a small area of the garden, using a fold unit or even an adapted rabbit hutch and run.

As bantams lay smaller eggs, you may prefer to allow two eggs each at breakfast (although 1 find one ample) and extra eggs for baking. Since 4/6 ordinary hens will supply the average family, you may need two or three extra if you have bantams.

What Age to Purchase

We have discussed rescuing chickens at around one or two years old, but not everyone will be confident enough to rescue chickens as their introduction to poultry. For the inexperienced keeper, the best age to buy a chicken is at the point of lay. This is usually around twenty weeks old when all the hard work of rearing, etc., is completed. The chicken, although at its most expensive, will then be ready to begin her working life. Although they are actually at the age when they will start laying, the upheaval of moving to a new home will delay the start for a week, maybe two. However, once they are settled, they will soon begin to produce eggs. You can purchase chickens at younger ages and they are usually cheaper.

Local breeders can be approached for younger stock. Young hens from eight to twenty weeks are termed *pullets*. The younger the hen, the cheaper she will be. If you buy them young, you are relieving the breeder of many weeks of feeding and housing them. However, you obviously then have to take over the task of continuing the rearing of the hen to the point of lay.

You should be able to find a local breeder to supply them; don't be tempted to order them by post. Day old chicks are virtually impossible to sex. This could mean you are actually purchasing cockerels as well as hens, so caution is required. The ratio could be as high as fifty per cent cockerels. You also need to obtain an electric or paraffin-powered brooder to keep the chicks warm. The brooder must be small enough to keep the chicks close to each other to provide extra warmth.

There is nothing nicer than rearing your own chicks once you have gained experience. If you purchase older chickens, you would be able to rear chicks at a later stage using your own eggs (Chapter 6, page 102). Of course, you would need a cockerel in order to make the eggs fertile.

How Many Chickens?

The amount of eggs you require will dictate the number of hens you keep. Four to six hens will keep an average family supplied with enough eggs with some left over for baking. Bearing in mind that a hen will lay five/six eggs a week, you can work out how many chickens you need. If you have a large extended family, a couple of extra hens would supply them also. You could ask extended family members for a contribution towards the layers feed if necessary.

During the moult, eggs will not be so abundant. However, if you store your surplus eggs as recommended (Chapter 5, page 75), then you should never go short.

Where to Buy Chickens

It is a good idea to ask locally for details of a reputable chicken breeder. Your corn merchant may know. Local farmers or even the local allotments will have a chicken keeper or two. The best place to purchase is one that can be recommended to you. If you use a recommended and reputable breeder, you can be certain that, if they say the chickens are vaccinated, then they will be. A breeder with a good reputation will want to keep it.

I would not advise buying from a market, particularly because of the threat of bird flu and the chance of other contagious diseases being present. If you buy from a market, you may not be able to rely on the reported age of the bird, etc. Has the bird been vaccinated? This would be difficult to establish. Buying any animal at a market is always a risky business, unless you really are an expert.

Mind you, having said that, I have purchased several older birds and a beautiful duck from a market. Although it was a pretty scary experience, it was a nice one too. It was rewarding to feel I was saving a few birds from an unknown fate and bringing them back to what I consider a good home. I also bought a batch of lovely quail that gave me many years of pleasure and some beautiful little eggs. However, I did not go to market until I had a little bit of experience under my belt, even then it was not really enough, and I think I was rather lucky with my purchases.

Transporting

Before you make the trip to pick up your stock, make sure everything is prepared. Food and water must be placed out ready to make the hens feel at home. The bedding must be laid and all sundries in place. This will allow you to unload the hens and leave them to settle in their new home.

Chickens do not seem to like travelling and find it very stressful. You must try to make the journey from the breeder to their new home a comfortable and painless one. Consider the weather conditions. If the weather is cold, then the container that they are travelling in needs air holes and insulating material such as a thick bed of straw or layers of newspaper. If the weather is hot, then provide more air holes and a cooling material such as shavings to travel on.

Buying special bird transport boxes is expensive and as a one-off is not necessary. A strong, sturdy box or crate is sufficient. If using a cardboard box, you need to line the base with a waterproof covering. A bin bag is ideal to protect it. (When transporting chickens they do tend to make rather a lot of droppings and this soaks the cardboard box causing the bottom to fall out.) Place your chosen bedding on top of the bin bag. As an extra precaution re-tape the bottom of the box to reinforce it.

If you have enough boxes, travel one hen per box. Or you can place several in one large box. When travelling together, the hens usually remain calm and quiet until reaching their destination. Tape the top of the box up securely, making sure the hens have no way out. If the weather is hot, drive with the car window open a little. This prevents the birds overheating. Do the journey as quickly as you can and do not be tempted to stop off at any other destination. The hens can overheat quickly in a vehicle during hot weather. Unload as soon as you return home. Place the chickens in their new home and leave to settle.

Fig. 12. Picking up a chicken

Not too much pressure around the ribs.

Handling Chickens

Place your hands either side of the chicken, closing them gently, making sure not to put too much pressure around her ribs (Fig. 12). Some breeds are more robust than others; the Warrens mentioned on page 45 are delicate little birds that could soon be crushed or hurt with too much pressure. However, they do like a firm contact. They can get rather upset if they are not used to being handled. If not held securely, they can flap their wings rapidly, causing them distress. Therefore it is vital to hold them with a firm but gentle contact. Make sure the wings are securely under the flat of your hand. If the hen feels secure, she remains calm and settled. If her wings are free to flap furiously, she upsets herself. I then place her either under my arm or close to my chest as in Fig. 13.

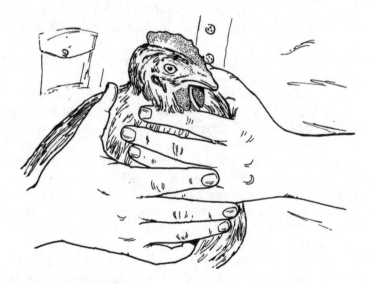

Fig. 13. Holding a chicken

Make sure she can't flap her wings.

If and when handling a broody hen with her chicks, you must take extra care. Once the broody hen has her chicks to look after, she becomes very protective towards the young-sters. Her mission in life is to stay alive and well to protect them. Everything is a threat to her liberty and her ability to protect the babies. I have found such hens become hysterical when picked up. Only handle the broody hen when abso-lutely necessary. Handle her in a similar manner to the normal hen, but make sure you have a slightly firmer hand. Be aware of her impending hysteria so that you don't get taken by surprise and drop her, etc. She will make a lot of noise but is not usually nasty.

The best time to handle chickens is when they are roost-ing. Chickens become very still at dusk and after dark. You can then pick them up without any struggle at all, allowing you to get a good contact around them. If you do go into the chicken house after dark, you can find the chicken you want by torchlight. Once you have found her, you can switch off the torch and take hold of her. Catching them in this way does reduce stress.

My personal view is that everyone should keep chickens. Children benefit from being introduced to the joys of chick-ens early on in life. Chickens give a small child an introduc-tion to responsibility and what is involved in caring for animals. We have addressed the issue of battery hens throughout this book. If everyone were to keep his or her own chickens, the entire battery hen industry would fail. Apologies to the reputable egg producers who keep their chickens free-range or deep litter, as I would wish no such fate upon them.

Whatever chickens you decide to work with, I am sure you will not be disappointed. There is so much fun to be had and so much enjoyment from eating your home-produced eggs. You can either go for the commercial layer, the Warren or some of the more decorative breeds. If these don't appeal to you, there are the little, friendly bantams. Basically, whether you are young or old, rich or poor, there will be a chicken out there that is ideal for you.

Chapter 4:

Feeding Your Chickens

The cost of feeding your chickens is your major outlay. It is often cheaper to buy in bulk if you can. The food needs to have a long sell-by date if it is to keep well. You must store the food in a cool, dry environment in order to keep to its finest condition. In a damp environment the food soon becomes dusty and mouldy.

Digestion in Chickens

Before we discuss what and when to feed, it is important to have an idea of the digestive mechanics of the chicken (Fig. 14). Birds do not have teeth and therefore cannot grind up the food with molars as most mammals do. However, they do have sharp projections on the tongue, which point backwards towards the throat. They help to draw the food to the back of the throat, enabling it to travel down to the stomach. The chicken, like all birds, has a unique thick-walled muscle beyond its stomach called the *gizzard*. The gizzard is the main aid to digestion of the bird's food.

The chicken eats the food, which is mixed with saliva from the salivary glands. From here the food will travel down the oesophagus (or gullet) into the *crop*. The crop is a temporary storage receptacle. When the bird has eaten, you will notice a large swelling usually to the side of the chest. This is the crop (Fig. 15). It is a good idea to take notice of the crop and its position on the chicken, just so that you are aware of it and you do not worry when you see swellings on

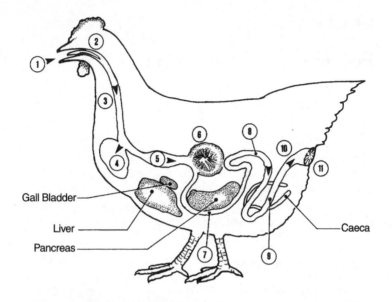

Fig. 14. Chicken's digestive system

Key:

1. Beak
2. Tongue
3. Gullet
4. Crop
5. Small stomach
6. Gizzard (grinds food)
7. Duodenum
8. Small intestine
9. Large intestine
10. Cloaca
11. Vent

the chests of your chickens. My first introduction to the crop was when I was tending my young chicks. I noticed one with a very large swelling and urgently prepared a trip to the vet. However, during the course of finding a box, putting the chick in and getting into the car, the swelling reduced. I

observed the chicken and during the next hour the lump
reduced even further. Thank goodness I never made it to the
vet's; that would have been rather embarrassing!

Fig. 15. The position of the crop

How very different a little chick looks when the crop is full!

The food, now softened, enters the stomach also known
as the *proventriculus*. Here the food is mixed with the digestive juices, hydrochloric acid and pepsin. The digestion
process is now under way; the food enters the gizzard. This
is a very thick-walled, muscular organ unique to birds.
Small stones that the bird has eaten help the gizzard to
grind the food up, ready for further digestion. This shows
the importance of grit and small stones in the diet of the
chicken. The food then travels down the duodenum and the
rest of the intestines, receiving digestive juices from the

pancreas and liver. The nutrients from the food can now be absorbed through the walls of the intestines. Undigested portions of food pass to the *cloaca*. In the cloaca the unutilised portions of food mix with the urinary waste products. From the cloaca the waste will be expelled through the *vent*.

Chickens do not pass urine like other animals; like all birds, they have no bladder. The kidney removes its waste in the form of a thick, white substance known as *urates*. This substance is expelled via the vent when the chicken passes a dropping. You will notice the urates as it surrounds the dark faeces.

The chickens we are discussing in this book are mostly layers. They therefore need a feed that will give the appropriate nutritional requirements for a laying hen. Layers pellets or layers mash is a complete feed that can be purchased from a reputable feed merchant. Layers feeds contain around sixteen per cent protein and extra calcium (for the egg shell), which is vital for the laying hen. A word of warning: quality is vital for the success of your layers. I have bought food from different suppliers and noticed that too much dust is present in some of the varieties. Having fed such feed I have then found my chickens reduce the number of eggs they lay.

Check the label on the bag to see that the ingredients are suitable for your layers. The list should include:

- Oil
- Protein (16% to 17%)
- Fibre
- Vitamins A, D, E.

You should also see listed trace minerals such as selenium. This is needed in a low dose in order to maintain good health. Methionine is an essential amino acid that needs to be present in the diet. This amino acid assists the breakdown of fat and aids towards an overall healthy digestive and urinary system. Copper sulphate is required in a low dose to help the chicken convert the feed into energy, etc., and is vital to life. The raw materials should also be listed. If the ingredients are not listed, then the food is not reputable!

It is best to fill a food hopper to just over half its capacity, thus giving the chicken access to food ad lib. If you are going to feed on a ration basis, then it is suggested as a guide to feed 3–6 oz (85–170g) of layers feed per chicken per day.

Never accept mouldy food from the merchant; mould has been known to kill chickens! Food that is below standard must be taken back to your merchant and changed or a refund given. Always check the dates on the bags and make sure you are not being sold old stock.

Food must not contain dust; if there is a lot of dust, then you must make sure that the supplier is aware. If you tip the food out into a bucket and dust flies everywhere, the feed is unsuitable and of poor quality!

If you do buy a complete feed, then you need not feed any other food with it. Feeding extra alongside the layers pellets or mash will cause an imbalance of vitamins and minerals. The vitamins and minerals are worked out and added in the correct proportions for the job. Adding other proprietary feed would also be a waste of money.

Always store your feed in a cool, dry environment; warmth or dampness shortens the life of the feed, causes mould to grow and the nutritional value to deteriorate quickly. Food should not really be stored for more than two months, so make sure you do not buy too much. A metal feed bin is ideal and keeps the temperature right. Store the food out of direct sunlight. Make sure a good lid is in place and always kept on. This keeps the food fresh and free from dust. It also serves to keep out vermin.

Feeding Extras

Chicken keepers worldwide have fed kitchen scraps for many years. Sadly, however, the practice of feeding household food waste has now become illegal. Since the outbreak of Foot and Mouth in 2001 certain precautions have been put in place by the Department for Environment, Food and Rural Affairs. The laws were introduced to prevent any future outbreak of disease caused by feeding waste that could possibly be contaminated with bacteria, viruses, etc.

In 2001 it was believed that feeding swill (food waste all mixed up into a swill) to pigs was the cause of the Foot and Mouth disease which caused devastation throughout the country. Salmonella in chickens is also likely when feeding scraps that possibly contain meat products. So now one cannot feed any form of catering waste to farm animals – and chickens are, of course, farm animals. This means that you always need to be sure that whatever extras you feed your chickens must not have been in contact with any meat products whatsoever.

As with all animals (including humans), boredom can lead to many behaviour problems. Bored chickens can resort to pecking their peers and eating eggs in the coop. In the past, chicken keepers relied on household food waste to add variety and stimulation to the lives of their chickens. We can still provide this but now it is vital that we purchase the vegetables from shops or markets specifically for our chickens. Vegetables can be purchased fairly cheaply from the local markets, and supermarkets often sell them at reduced prices. At least with this method you will probably have larger quantities. You will also find that, as the vegetables are specially purchased, the chickens will have better quality food as opposed to scraps, which are, of course, leftovers.

An ideal ration of fresh vegetables would be in the region of 1 oz (28.35g) per bird; much above this amount and they would not consume sufficient layers rations to give an abundant supply of eggs. As already mentioned, layers pellets and mash are complete, and too much food alongside would upset the balance provided. The better fed the chicken, the better her laying capacity and the better the quality of the eggshell.

Vegetables can be tied up in the chicken house to provide fun and stimulation. The chickens will enjoy many hours pecking at them. You can purchase any vegetables for this procedure. Carrots, turnips, parsnips and Swedes are ideal as they are easy to hang up (vegetables can be lightly cooked to soften if required). Simply make a hole in the middle of the vegetables and thread string through them. They can then be hung from beams or doorframes.

Green and yellow vegetables, such as corn on the cob, are rich in vitamin A. This darkens the egg yolk, making it a lovely deep yellow. Potatoes are enjoyed enormously by the chickens. If the weather turns cold, chickens appreciate a pan of boiled up potatoes mixed with layers mash (this is best served warm in winter). This is a very hearty and warming treat for your birds. Never feed raw potatoes. Always cook them to a nice soft texture; your chickens will love them. If you have electricity outside, you could always cook up the potatoes in an old slow cooker!

I have noticed that my chickens love oranges. If you are really lucky, these can be bought very cheap late in the day at markets.

Bread contains calcium and protein and also allows for you, the keeper, to interact with the chickens while feeding the bread. I often sit out in the sunshine and break up the bread into small pieces, throwing it to my chickens. They also like it placed in a bowl of water thus making it a soggy treat.

Mixing Your Own Rations

Some chicken keepers like to mix their own rations; this way they know exactly what the chickens are eating and the quality of the food. Mixing also allows the keeper to be in control of any additives.

You would have to enquire locally to see if your own feed supplier would offer you a mixing service. These days convenience seems to win over and very few suppliers offer this service. Word of mouth is a good source to find out if there may be a local farmer who has his own mill; if you are lucky, it may be made available for you. Failing that, you might be fortunate enough to get your hands on a hand grinder.

An alternative would be to compile a list of your chosen ingredients and rations. You could then purchase each ingredient separately and combine it yourself. This is harder work but it must add variety into the lives of the chickens. This method would suit free-range chickens that have access to grit and small stones. You may find that the indoor chicken may have difficulty grinding up whole grains.

A simple mix for laying hens would consist of:

- Bran
- Wheat meal
- Ground oats
- Maize meal
- Meat or Fish meal (add only half portion of this).

An equal amount of each of the first four ingredients, i.e., a large scoop of each, should be measured out into a large container and mixed well. Adding some water makes the meal more palatable. Barley could be added but not in large amounts. Remember: barley will fatten the bird, thus reducing egg production. Adding dried milk would enhance the calcium content.

A mixture of scratch corn can be given in the evening when whole oats can be given, along with whole wheat and split maize. Again, the mixture can be made with a large scoop of each, all mixed into a large container and scattered at teatime. This could also be a good way to encourage the hens into the coop at night.

A good additive is sprouted grain. Place quality cereal, such as oats, in a bucket and fill to the top with water. Leave for a minimum of twelve hours and then drain off the liquid. Cover the bucket and place in a warm, dark environment; stir the contents of the bucket two or three times a day. The sprouts will grow rapidly and can be fed when half an inch (1cm) long. These are ideal, fed in the winter months. They add nutrition and variety.

Vital Food Contents

Carbohydrates make up a very high percentage of a chicken's diet, as is the case with most animals. They play an important role in the maintenance of body tissues and provide a lot of energy. Chickens use a lot of energy in scratching around, foraging, preening, mating and, of course, laying eggs.

Fats and oils provide energy and warmth for chickens and can be supplied in many forms. Peanuts are a good source. Chickens seem to love margarine and butter which are an excellent source of fats and oils, especially in colder weather. I occasionally indulge my chickens with a cheap tub of margarine from the supermarket.

Proteins are vital to repair and build tissue in the body. Protein can be obtained by providing fish and blood meal. Left-over milk also provides some of the protein required. Protein molecules are made up of amino acids of which there are twenty-two. Eight essential amino acids can only be obtained by eating animal tissue or plants. Worms provide good solid protein; grains and beans are a good source also.

Calcium and phosphorus, along with the grit, are provided in the form of oyster shell. Limestone flour is a good source of calcium and can be served in a container, allowing the chickens to help themselves.

Any laying bird or growing chick needs a high protein diet. No single food source can provide all the amino acids required, therefore a good feed consists of many different sources. The main sources should consist of meat, fish and bone meal, along with plant proteins such as soya beans and cottonseed meal.

Vitamins

All good purchased food contains the vitamins required for a healthy diet. Check the label on the bag; this gives full details of the contents. Sprouted grains provide a good source of vitamins. Vitamin D is provided for free-range birds by the sunlight. Vitamin D must be present to help with the uptake of calcium and phosphorus. Bones are strong and well formed when the birds have plenty of vitamin D. If the calcium and phosphorus are utilised correctly, the eggs will be laid with a good shell. If the chicken is not free-range, then vitamin D must be provided. Fish meal and cod liver oil will give this much-required vitamin.

Vitamin A helps the chicken to retain a good, healthy skin and will lead to a healthy gut, healthy respiratory tract and healthy reproductive organs. Vitamin A also helps to keep the chicken free from disease. Yellow maize is high in A, as is grass meal.

Supplements

Visit any good, reputable feed store and you should have access to many supplements and additives. There are some really good purchases out there but it is wise not to get carried away. If you have adhered to the feeding ideas put forward in this book, you should have little need for additives, etc. However, I feel it is important that you should know about what's available.

Cider vinegar is a natural product. It is a tonic, an antibacterial and an anti-coccidial product that can help towards the health of all poultry. Egg supply should be improved and the chicken should take on a healthier look about its feathers. The vinegar is easy to administer as it is simply added to the drinking water of the chickens.

Game Guard is a probiotic that contains lactic acid-forming bacteria: *Lactobacillus acidophilus* and *Enterococcus faecium*. Use this product at times of stress, such as moving house or after hatching.

Intervits Soluble Multivitamins: this product also contains a water-cleansing agent. A vitamin supplement, it is useful if you feel that your birds are missing out on some important nutrition.

Vetark Zolcal-D helps with eggshell production and is high in calcium. Suitable for laying chickens as it encourages maximum calcium consumption.

Shell Max is a soluble calcium/D3 supplement containing minerals that improve shell strength through the laying period.

Cod liver oil is a good supplement for all livestock. This product improves just about everything. The chicken improves in overall well-being and physical appearance. The shell improves in quality and the chicken makes a speedy recovery after the moult period. The chicken utilises the calcium in her diet more efficiently. Extra vitamin A and D are also provided.

Water

Water is a vital element of chicken husbandry. It is extremely important to provide plenty of fresh, clean water. Water containers should be cleaned daily; water should be neither too hot nor too cold. I find that most of my animals like the trough water, which I use for all the poultry, horses and cows.

If your chickens are free-range, you will find that they will seek out water in the most extraordinary places. They seem to like to wander around and try different varieties. Puddles, pools and any old rainwater caught in flowerpots and containers can provide chickens with a little variety. I have often found my chickens drinking the most disgusting-looking pool of water. There must be different flavours of water that the chickens will enjoy. I like to leave a few containers around the yard to catch a supply of rainwater.

It is important to remember that just a couple of hours deprived of water can cause severe dehydration. Without water, the chickens tend not to eat the corn, so with a combination of the two deprivations your egg production will be affected greatly. Remember that sixty/seventy per cent of the egg is made up of water, hence the reduction in egg production during water deprivation. So check that the water supply is functioning correctly all the time.

Water helps to remove waste products produced by food and exercise. As food is eaten, water softens the corn or pellets and aids the extraction of vitamins and minerals from the food. Chickens, being small animals, do not ingest great amounts of water at a time, however they do make many visits to the water supply throughout the day. Many of the supplements available (some mentioned above) are administered by adding to the water.

Calcium and Grit

If the chickens are short of grit, they do not get the maximum nutritional capacity out of their rations. Like all birds, a chicken's digestive system relies on the gizzard. The gizzard needs grit in order to function correctly. Free-ranging birds usually manage to peck around and find grit in the soil. But if the chickens do not have access to grit, then it must be provided for them in a container. Grit can be obtained from the feed merchant and is not expensive. Once purchased it will last a long time.

Calcium is another important element in the diet of the chicken. The shell of the egg is made up of inedible calcium; without calcium in the diet of the chicken, the shells do not form properly. The shells are soft and break easily as you collect them up. I have known eggshells to be almost paper-thin when calcium is lacking. Eggs can also be laid without any shell at all which will encourage the chickens to eat them. Once the chickens get a taste for the eggs you have a major problem in the coop! The general health of the chicken also suffers and skeletal problems may occur. Chickens can become very weak; egg production is also put at risk. A chicken requires extra calcium two weeks before she is due to start laying eggs. Chickens usually lay at about twenty-two weeks. Start to feed extra calcium at around nineteen weeks. Tests have shown that chickens require a hundred times more calcium when laying eggs than is required when off lay!

Calcium can be obtained in the form of oyster shell, which contains not only calcium but phosphorus as well. It also provides the grit requirements. Adding limestone to the diet can also provide calcium. Another little tip to help with calcium is to recycle all your used eggshells. The eggshells are baked in the oven on a low heat until they are a crumbly consistency. You then need to crush them until tiny, as the chicken must not know she is eating eggshell. If the shell is left in big pieces, she might recognise it and begin eating eggs as she lays them.

Control of Vermin

Vermin must be controlled. There is an old saying, "If you have chickens you have rats." They breed and thrive on your chicken feed. Cats are a good method of control, one of the best ways of keeping on top of the numbers of rats so that they do not take over. A good rat-catching cat will be worth its weight in gold! If you do use cats, then they must be fed and watered well. It is a myth that cats will survive purely on what they hunt. They don't. A cat can only do the job of hunting if he/she is kept fully fit. The cat will also need regular worming due to the nature of the work.

Rats can reproduce from the age of four months. The gestation period is around twenty-three days. The female rat can produce upwards of five litters per year with ten babies per litter. Considering again the fact that each of the fifty young produced will start breeding at four months, you can see from the arithmetic how quickly a serious rat problem can occur! I know from experience that the rats can end up eating more of the feed than the chickens.

Poison left for the rats can be dangerous for other animals such as dogs and cats. It can cause illness or even death. I find that the poisoning of rats is rather inhumane; the rats can take a long time to die, which I always find distressing.

It is still a difficult issue for me as I worry about the suffering of any living thing. Voice your rat problem when you are out and about. Someone who enjoys shooting may get to hear of your dilemma. The gun is instantaneous and suffering is minimal.

Rats have an amazing ability to gnaw through the structures of buildings and any surrounding materials. They leave behind large piles of material that they have gnawed away at overnight. They seem to go at anything that stands in their way and can cause extensive damage. They may chew their way through your electrics, thus causing a fire hazard or breakdowns. They also seem to have a talent for digging up flooring to make their tunnels.

Rats are known to spread over thirty diseases to humans and many diseases to the animal population. One of the most worrying diseases is *leptospirosis*, which is spread via the rat's urine. Salmonellosis, rabies and ringworm are others, to mention but a few. Rats can also transport mites and fleas, causing massive infestation in your chicken house.

A Little Bit About Rats

The roof rat (also known as black rat though he may well be brown in colour) is a rather fussy eater and prefers cereals and grains. Roof rats are now very rare in most parts of the UK. They also like fruit and sweet foods such as cake, which is different from the much commoner Norway rat, also known as the brown rat; he likes fresh meat and fish and is partial to a bit of fruit too. Rats tend to come out mostly at night, although they can sometimes be seen in the day.

Rats are very able bodied and agile; they can shimmy up a vertical wall at great speed and can also cross thin, high structures such as wires. They can leap on and off feeding hoppers with great agility even when these are placed rather high. Of course, you can never put the feed hoppers any higher than the chickens can reach, which puts a limit on the height available.

Poison

Sadly for some people, rats can only be controlled by the use of poison. Before trying this, it is a good idea to leave out food for the rats and make sure they like it. You could put food out in one of their paths and see whether it is eaten. You may have to test different varieties of food. Fruit, cake and meat can be left out in turn and then you will know which they prefer. Once you have established the food they are most partial to, you can begin to administer the poison on the chosen bait.

Make sure that other animals and children are unable to have access to the poison, be it out on bait or in its container. Poison should always be locked away where it cannot be stumbled upon by chance.

Many of the poisons on offer for rats these days make cats and dogs poorly, so they will expel the poison by being sick. Dispose of dead rats as soon as you are aware of them. This will help prevent other animals eating the bodies. Always wear gloves when picking up dead rats.

If you are using poison in an occupied hen house, you must put the poisoned bait into a baiting station. You can buy baiting stations like the one shown in Fig. 16, or you can make your own. An easy but effective method is a large, wooden box turned upside down; a small square or hole is cut into the box, leaving just enough space for the rats to enter but small enough to prevent entry by your chickens. The box should be sturdy enough to stop them knocking it over.

Rats are often suspicious of anything new placed in their territory and so the bait, when placed, will not be touched straightaway. It may be several days before they start to take it.

Anti-coagulants are the main rodenticide used for the control of rats. Anti-coagulants interrupt the function of vitamin K, which helps blood to clot. When the blood fails to clot, haemorrhaging (bleeding) will occur. Vitamin K is often present in chicken corn and therefore the rats should not have access to it, if at all possible.

For rats that are genetically resistant to the anti-coagulants available to poultry keepers, there are alternatives. Only professional pest controllers can supply these,

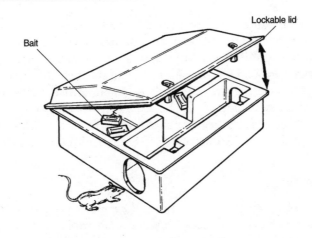

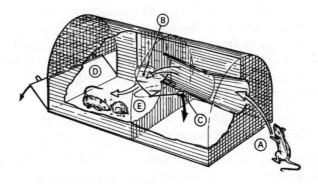

Fig. 16. Dealing with rats

(top)   A baiting station. The top closes securely so that children or other animals can't get at the poison.

(bottom)   A humane rat trap. When caught, the rat can be humanely shot or taken a long way away and released. Rat enters at A, runs along cage and down through trapdoor B (which is closed by a counterweight C) thus entering the secure baited enclosure E. Later, the rat is removed through door D.

which include *brodifacoum* and *flocoumafen*.

Most rodenticides can be purchased in a ready-to-use condition. Whatever you decide to purchase, you must make sure you follow the instructions carefully. It is also vital that all other livestock is taken into consideration. Even more important is the storage of the poisons; I make no apology for repeating that poison must be locked away in a safe place, unobtainable to children in particular!

Rat Traps

Trapping can be carried out on the smaller poultry unit but is impractical on larger farms. Any tasty morsels can be used as bait with the traps placed close to the walls, in paths that the rats normally take. Rats do tend to have visible paths that they follow strictly. The rats' bodies must be removed on a very regular basis to prevent flies and maggots.

Ultrasonic plug-in devices are available. The idea is that only the rats can hear the high-pitched ultrasound which will drive them away. I have tried these with negligible results; however I have friends who recommend them.

To Summarise on Feeding

Feeding your chickens well will reap its own rewards. Supplementation should not really be required if you use a reputable feed merchant. Vegetables are, however, an important addition to the life of the chicken. They add variety and relieve boredom. Greens and sunlight should provide plenty of vitamins. Grit will help the chicken utilise the food. Just make sure that while providing your chickens with a good diet you are not providing for the rats as well!

Chapter 5:

Care and Use of Your Eggs

Egg Collection

The collection of your first freshly laid eggs is a momentous occasion. Do not underestimate the great feeling of achievement and joy as you peep into the nest boxes for the treasure. I know that you have not laid the eggs yourself, but you have certainly helped towards them.

Always take great care when dealing with the eggs, as they are very delicate. You will need a suitable basket, box or bucket, depending on how many eggs your hens are producing. The container needs to be lined with a material that will cushion the eggs and keep them safe. Hay, straw, shavings or shredded paper are ideal; place the eggs gently into your chosen container and handle with care!

If you purchased your chickens at point of lay, you may have to wait a week or so for them to start laying. The move to new premises will cause a delay in the start of production. Once laying begins, you will find that the number and size of the eggs will vary initially. Once into the swing of things, the eggs will become a more regular size and the numbers will be more predictable.

The breed of your hens will determine the colour of the eggshell. Chickens with white ear lobes will lay white eggs. Chickens with red ear lobes will lay brown eggs. There are also chickens that produce blue eggs and speckled eggs. You may sometimes be lucky enough to find double yolkers in your nest boxes. These are basically twins, two for the price of one. I find it thrilling when I crack a shell and out pop two yolks. Double yolkers often are very big eggs; some have a tell-tale ridge around the middle of the shell. I find the ridge generally to be a good indication of a double yolker. I always ensure I have an audience when I crack them open.

Place the eggs in your basket/bucket close to one another so that they do not roll around and crack on impact with another egg. I usually check the nest boxes when I turn out my chickens first thing in a morning. However, you will need to return to the nest boxes regularly during the day as eggs will be laid at various times. Hens usually lay once a day but it is not always at dawn. I have known many chickens lay after tea or early evening. It takes just over one day (25–26 hours) for the egg to be laid. This is the time the egg takes from ovulation to the actual laying. This is why eggs are laid later in the day as the week progresses. So, on a twenty-four hour clock it is impossible for the hen to lay at the same time each day. She will lay for several consecutive days and then have a day of rest to begin the cycle again.

If you are working during the day, then I suggest that you collect before work and on your return home. If you do not make regular collections, your eggs may become damaged. It only takes one broken egg for the chickens to realise how yummy the contents are. Chickens may then resort to breaking the eggs themselves to get at the contents. Once this problem is allowed to develop, it is hard to discourage. If a serious problem with egg eating does occur, you will need to find the culprit quickly. Remove her from the flock to an area of her own. Remove her at dusk when she is roosting and place her in a safe covered run or hutch. Fill some eggs with mustard by putting a hole in each end of the egg. One hole should be bigger than the other so that you can blow into the small hole, thus pushing the contents out of the egg through the larger hole. Then use a syringe to fill with mustard and cover the second hole with tape. Leaving the mustard-filled eggs with the hen should break her of the habit. If you cannot find the culprit, place some mustard-filled eggs in the actual nest boxes. A mustard-covered beak gives the perpetrator away! If you fail to break the habit, remove the guilty party to her own run or cage, watch her closely and remove her egg each day before she can eat it.

If your chickens are free-ranging over a large area, you may discover they start laying outside, as opposed to inside. Try as you might to find the eggs, it is not easy. I notice that I only find nests when I am not looking for them. As I am going about my daily tasks around the farm I often stumble by chance upon a well-hidden nest of secret eggs.

If you do find your egg production drops rapidly (provided disease is not the cause), your chickens may be laying when they are let out. Sometimes, if you observe your chickens as they come out, you may see the culprits running urgently to their hoard of eggs. If you watch the direction they run to, you will probably be able to find your missing eggs.

Upon finding a nest, gather up the eggs and keep them separate in case they are older eggs. I often leave one egg so that they continue to lay in this same place. That way I know where to find the eggs. Crack these eggs into a cup before use to make sure they are fresh.

Storing Eggs

The storage of the eggs is very important, as you need to make sure that they are kept at the correct temperature to maintain the quality of the egg. A suggested temperature is 40 degrees Fahrenheit or 4 degrees Celsius. Too much above this temperature could cause the egg to lose some of its nutrition. Eggs need to be stored with the rounded end facing upwards. The pointed end needs to face into the egg box cups. Storing eggs in this position allows the yolk to sit well in its shell. The yolk is also suspended with maximum protection from the white, which contains anti-bacterial properties. If you are storing the eggs for any length of time, keep the trays at a 45° angle and rotate them through 90° once a week. This is done by lifting up the tray at the other end each time. Turning must take place daily if you are thinking of hatching any of the eggs off (see page 89).

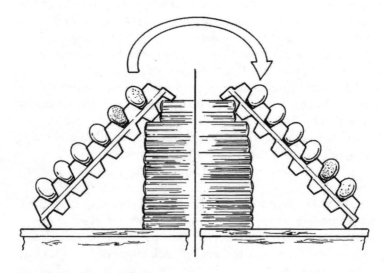

Fig. 17. Rotating the eggs

Rotate them through 90° by once a week changing over the end that is lifted up.

I have found that many people like to buy dirty eggs when I am selling them. People seem to like to see the dirty shell, as they associate this with a hen which enjoys life in a non-sterile environment. They particularly like to have a feather in the box, which adds a magical feel for the older person in particular. I would advise you not to clean the eggs until the actual moment you are going to use them. Eggs are laid with a protective coating (bloom) and this can be easily washed away, thus allowing bacteria to enter through the porous shell. However, badly soiled eggs must be cleaned when collected to prevent contamination from germs. Clean the eggs in warm water, wiping with a soft sponge or cloth. You can buy special egg washing powder or detergent to clean the eggs if you feel this is necessary.

Eggs work better stored at room temperature; however, they will keep longer when refrigerated. If you do put them in the fridge, don't be tempted to store them in the fridge door. The temperature will fluctuate a great deal each time the door is opened when you are in and out for milk, bacon, etc. You are better off putting them in an egg box inside the fridge. In the fridge eggs should keep for up to five weeks. Free-range eggs often keep for a little less time due to the less sterile conditions they are often laid in. That is why it is wise to keep the nest boxes as clean and well bedded as possible.

If you are lucky enough to have one, a cellar is an ideal place to store eggs, as it is dark and cool. Failing that, any cool room will do; a room where the temperature is constant no matter what the weather is outside.

If you only have a few hens, you will not have much trouble using up surplus eggs. If you do get a large number of surplus eggs, then you must use them in rotation. The oldest eggs must, of course, be used first. You could consider getting a little ink and stamp set from the stationer. Or you could possibly mark them lightly with a felt tip pen. Only mark lightly to prevent the ink penetrating the eggshell.

Freezing Eggs

Eggs can be frozen, provided the correct method is used. In the freezer, eggs can keep for up to six months. There are three methods of freezing your eggs:

- Freeze as a whole egg

- Freeze as just egg white
- Freeze as just the yolks.

To freeze the egg whole, you need to crack the egg into the chosen freezer-proof container. Give it a mix with a fork to combine the white and yolk. You can freeze several eggs together to be used in suitable recipes. If you make a lot of sponge cakes, you could freeze, say, four eggs to be used in a cake mix. Label the number of eggs in the container and the date frozen. These eggs can than be thawed out and used in cakes or omelettes.

To freeze yolks, you can separate them from the white and freeze in ice cube trays, which are a handy size, or you could combine several in a freezer-proof container. The same applies for the egg whites, which can be frozen together or as separate whites. Remember to label and date. To prevent egg yolks becoming too thick as a result of freezing, add a pinch of salt. Label the egg as salted so that it is used in savoury recipes. For sweet recipes several yolks should be frozen with half a teaspoon of sugar added to prevent thickening.

Preserving Eggs

Water glass (sodium silicate) should be available at the local chemist. Use the water glass in a ration of one part water glass to ten parts water. Water glass should be added to boiled (hot) water and mixed well. When the liquid glass has cooled, you can then dip the eggs in the solution. The size of the container used depends on the number of eggs to be preserved. Do not wash the eggs before the preservation; they will keep better when their natural coating is intact. You can use the eggs from the container when required. Just bring them out and wash the solution away. Eggs preserved in liquid glass can keep for up to a year.

An old method from many years ago was to place the eggs

in ash from the fire (cooled, of course). The idea was to place them gently in a container of the ash and keep them in the cellar. Of course, cellars are not commonplace these days so you would need to place them in a similar cool and dark environment.

Another method from yesteryear was to melt lard, and coat the eggs in the melted solution. The eggs, once coated, were placed on a dry surface such as newspaper, greaseproof or blotting paper to dry out. The eggs were then carefully layered in a flat box until they were required.

Another suggestion is to pack in salt or rolled oats. This would produce a constant cool and dry environment to keep the eggs fresh.

Sometimes, when cooking a specific recipe, you may find that only one part of the egg is required. Meringues, for example, only require the white of the egg. In such circumstances, you can store the redundant part of the egg in the fridge for up to four days. The egg should be placed in the fridge in an airtight container. Yolks remain at their best for only two days, and they are best stored with a covering of water to stop the yolk drying out.

Pickling Eggs

Pickled eggs are very popular with salad or when having a beer (apparently). They are also very nice with cold meat or cheese. As with any form of preservation, pick the finest and newest eggs of the batch. Boil the eggs until hard in the centre. Cool under the cold-water tap so that you can pickle them straightaway. You get maximum quality and nutrition if you preserve them quickly. Remove the shell; this may be difficult with fresh eggs. I always find when trying to peel a fresh egg that I can only pick bits off at a time. Don't worry; this is just another indication of their freshness.

Once they are peeled, place them gently in your pickling jar. Fill the jar with eggs but make sure you don't over fill and squash any of them. Boil the pickling vinegar with a half ounce (14g) of peppercorns and pour it into the jar until all the eggs are covered. Seal the jar tightly until the eggs are required.

Good or Bad Eggs

Free-range eggs are delightful when fresh from the nest. However, you must always take care that the quality of the shell is good enough to keep the egg fresh inside. Some eggshells will be thin and almost transparent; you can tell by the texture that such a shell will not keep bacteria out. Always use these eggs first. If you store this type of egg, it will go rancid very quickly.

Another problem to look out for is a hairline crack in the eggshell. Sometimes eggs are damaged in the nest if several chickens are in and out laying during the morning. You may not be aware of the thin crack and the egg will go off very quickly. When placing your eggs in the basket, take note of the sound they make when touching another egg. When you get used to the sound as they gently touch, you will notice when you put a cracked one in. Though you may not see the crack, the sound of shell touching shell is different when one of them is cracked. It is difficult to put into words; you will know it when you hear it. The only way to describe it is a dull and hollow sound when an egg is cracked. When the eggs are fully intact, the sound is one of a more rounded, tuneful note.

For the above reasons it is sensible for cooks to crack all eggs one at a time into a cup before using them. This is a particularly good idea when baking cakes. I have known a friend ruin a whole cake mixture by putting a rancid egg in at the end. When you think of the sugar and butter used in a cake mix, that is a lot of waste.

A fresh egg in the frying pan should have a lovely orange or yellow yolk with a thick white surrounding it. If the egg white runs all over the pan, you know you have an older egg or possibly a vitamin deficiency in the diet of the chickens!

It is handy to know, before you crack open an egg, if it is fresh or not. Fill up a bowl with cold water and carefully place your egg in the water. Support the egg so that it does not fall rapidly to the bottom of the bowl. If the egg sinks to the bottom, it is fresh. Should the egg bob about on the top of the water, you can be sure it is a bad one. The floating is caused by the enlarged air sac. Every egg contains a small air sac; as the egg ages, it loses some of its moisture content. As the moisture lessens, the air sac becomes bigger. If the egg floats like a rubber dinghy, you must dispose of it very carefully. Should you not take care, you could end up with egg on your face! If the egg touches the bottom but bobs slightly upwards, then it is slightly older but will not be bad.

Candling is also a good method to test for freshness (Fig. 18). A special electric candling lamp is used for this purpose (the same method is used when planning to hatch eggs, see page 98). The candling is carried out in a dark room. The candling lamp is switched on and the egg held onto the light. You must place the blunt end onto the light. The lamp will show up any blood spots or meat spots. It will also reveal how large the air sac is. The air sac will begin life very small. As the egg ages, the sac will increase in size.

If you just want to carry out a quick assessment of the quality of your egg, simply shake it vigorously back and forth closely past your ear. If you can hear the egg moving around in liquid, be cautious. The more liquid you can hear, the staler the egg. This method takes a little practice but I have found it generally valuable. On odd occasions I have passed the egg by my ear and heard nothing, yet still the egg has been bad. So you cannot depend on this method entirely! Once you have got used to the variations in the sound, you will usually know when you have a bad egg. With a very watery sound the same rule applies: dispose of the egg very carefully!

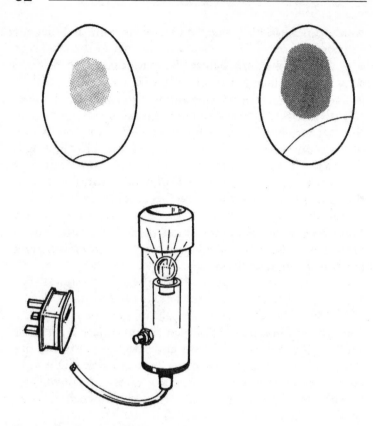

Fig. 18. Candling eggs for freshness

Place the rounded end onto the light.

Using Your Eggs

I would like to mention here how to fry an egg. It may sound silly but not everyone can fry an egg properly. Once you start to cook with your home-produced eggs then it is no longer the same as just frying an egg. It becomes an art form!

- Heat vegetable oil in the frying pan until a haze is present
- Remove the pan from the heat, wait a minute or two
- Turn the heat down on the hob or ring
- Gently crack open your fresh egg and lower the contents carefully into the base of the pan
- Place the lid on the frying pan
- Return the frying pan to the heat, which is now very low
- Wait until the yolk has an opaque film over the top
- Serve.

This method should produce the perfect fried egg with a lovely soft yolk. If you like a harder yolk, then simply cook for a couple more seconds.

Boiled Eggs: Soft

Place the egg in a pan of boiling water and time it for three and a half minutes, depending on preference. I prefer five minutes to be sure it is cooked. A friend of mine did inform me that exactly six minutes produces the perfect boiled egg. It depends a bit on the size of the egg. Bigger ones take longer.

Boiled Eggs: Hard

Place the egg in boiling water and simmer for ten minutes.

Poached Eggs

The special egg poaching pans give the best results and are easy to use. If you do not have such a pan, then boil an ordinary saucepan or frying pan of water. When the water is boiling, lower the heat to simmer. Crack an egg and gently slide it in. Simmer until the desired texture is reached. This method can be rather hit and miss. You can either get a really nice egg or a rather messy one.

Another method is to boil the egg in its shell for twenty seconds to help the egg retain its shape. Remove the egg from the water and gently crack it open back into the pan. Boiling it in the shell for a few moments sets the white a little. This should give you a better-shaped and set poached egg. Cook to the required texture.

Scrambled Eggs

For each egg use a tablespoon of milk and a knob of butter. Mix the eggs together and add the milk. Add salt and pepper to taste; melt the butter in the pan. Add the egg mixture and stir while on a low heat until the mixture is set to a fairly soft consistency.

Omelette

Whisk three eggs thoroughly with a tablespoon of cold water, season with salt and pepper. Heat a well-greased pan and place the mixture in. Keep loosening the edge of the omelette by running a knife around the edge. Cook gently until the omelette sets, fold it in half and serve. You can fill the omelette if you wish with your favourite filling.

Recipes

An obvious use for surplus eggs is baking cakes. There are many recipes for a sponge cake, be it a fatless sponge or a Victoria sponge. If I have left-over eggs or cracked, damaged eggs, I will always make a sponge or two. Sponge cakes can be put in the freezer to use at a later date.

Victoria Sponge

This recipe keeps for three or four days in an airtight container, and the amounts given here make two or three

Victoria sponges, or could make a couple of cakes and a sponge pudding. Surplus cakes can be frozen for eating later. It's a good way of using up cracked or broken eggs which shouldn't be kept hanging around too long.

Start with 10 oz (284g) butter, 10 oz (284g) flour and 5 eggs. Whisk the butter, sugar and eggs until combined (you may need to add a tablespoon of flour to help the ingredients to combine). Then fold the rest of the flour in gently and place the mixture in an oiled cake tin. Cook on a medium heat until golden brown (about 20-25 minutes). If you think the cake is done, hold it up to your ear. If you hear crackling, the cake needs a little longer; if silent it is ready.

Eggy Bread (sometimes called Gypsy Toast)

This recipe makes a nutritious and nourishing breakfast or tea. We used to have this a lot when I was little. You can either make this as a savoury or as a sweet. Use one egg to each slice of bread; whisk the eggs well. You can add a dessertspoonful of milk per egg if you desire. Add salt to this mixture for a savoury dish!

Soak a slice of bread in the egg mixture and fry until golden. If you are having a savoury dish, serve with tomato sauce or similar. If you are having it as a sweet, you can toss it in sugar – delicious.

Meringues and Egg Custard

I mention these two together as one uses the yolk and the other uses the white. Meringues are so easy to make and so impressive to serve to family and guests. Meringues whisk better by using older eggs at room temperature. Fresh egg whites are far too tight for the recipe. So not only can you use surplus eggs but older ones also. When doing the water test, use eggs that just start to rise off the bottom of the bowl. Make sure all utensils used are clean and free of grease.

Use the egg whites only for meringues. Gently crack the eggshell in half and send the yolk back and forth between each half of the shell. As you do this, allow the white to drain away into a bowl. Each egg white will require 2 oz (57g) of caster sugar. I rarely have caster sugar in but find the meringues just as lovely with granulated sugar.

Whisk the egg white until the mixture forms soft peaks. You will know it is sufficiently mixed when you can hold the bowl upside down without any loss of mixture. Add the sugar gradually whilst still whisking.

Grease a baking sheet and put greaseproof paper on top. Place a dessertspoon of the mixture on the baking sheet. Make sure you have an odd number of meringues so that you can eat one when they come out of the oven! Leave in the oven on a very low heat to dry the meringues. Two/three hours should be ample time. When you think they are done, switch the oven off and leave to cool. Fill with fresh whipped cream and serve.

Egg Custard

You can use whole eggs but it is handy to use up egg yolks as the dish cooks the same. Beat together 4 egg yolks and 1 pint (568ml) of milk; stir in 1 oz (28g) sugar. Place in a two-pint bowl then sprinkle with grated nutmeg. Stand the bowl in a baking dish with water half way up the outside of the bowl. Bake in the oven on a medium heat for just under an hour. Alternatively, the egg custard mixture can be placed in a lightly cooked pastry case before putting in the oven.

Pancake Mix

Using 3 oz (85g) of flour, sift with a pinch of salt. Make a well and place in it 3 eggs. Beat the mixture well while adding half a pint (284ml) of milk slowly. Leave to stand at room temperature for about an hour before use. Heat a drop of oil in a pan and add a ladle of the pancake batter. Cook gently, turning a couple of times until golden brown. Serve with your favourite filling.

Scotch Eggs

I just have to mention Scotch eggs! This is my favourite way to use surplus eggs. Once you have tasted a home-made Scotch egg it is not easy to eat shop-bought again. These are simple to make yet give a very impressive result. Purchase 1 lb (450g) good-quality sausage meat and a box of bread-crumbs. Hard-boil half a dozen eggs; cool and peel. Divide the sausage meat into six; roll out each portion enough to cover the egg. You can sprinkle the inside of the meat with a little pepper or something that you really like; just be adventurous. Mould the meat around the egg, roll it in a plate of raw egg and then roll it in the breadcrumbs until covered. When you have covered all the eggs you can then deep-fry them until crispy golden. Serve with salad, chips or just as a snack (I didn't say this was a healthy recipe!).

Eggs for Beauty Treatments

I can remember using egg beauty treatments when I was younger. I would save unused egg whites and coat my face with them. It would dry, leaving my skin feeling very tight, and then I would wash it off. I must say it always left my face feeling very fresh and smooth.

You can also use yolks on the skin. Rub in a massaging motion into the skin on your face. The massaging and gentle rubbing will remove all dead skin and other impurities. The skin will be left feeling fresh, soft and vibrant. Mind you, I must admit I did not like the feel of rubbing egg yolk into my skin.

The yolks also make a nice hair conditioner with the

added vitamin content penetrating the shaft of the hair and the follicles, etc. You just have to be careful not to use very hot water when rinsing. Over-hot water could leave you with a scalp covered in omelette! Rub the yolk well into the hair and scalp after washing; rinse thoroughly after ten minutes.

If you were planning to go out for the evening and found yourself devoid of all hair setting products, you could always use a mousse substitute. Whisk up an egg white and use it as you would use hair styling mousse.

This whisked-up egg white mixture can also be placed on stubborn stains. Leave the mousse on a stained item of clothing for ten minutes then remove and wash as normal.

So, as you can see, you never need to waste an egg. If you look after your eggs, they will look after you. Keeping your own chickens means you never need to go hungry nor to go out looking anything other than beautiful!

Chapter 6:

Extra Joys of Chickens

During my time as a chicken enthusiast and keeper I have had some real high points. One of many has been the discovery of the broody hen! She is a joy to behold and brings real meaning to the old and well-used expression of 'taking someone under your wing'. I had been led to believe over the years that a broody hen was something to be shunned and scorned. Not so! I find the broody hen to be the epitome of motherhood, someone we could all learn from.

So don't be put off when stumbling upon this condition in the hen house. Embrace the broody hen and allow her the joy of motherhood that she so desires. Believe me, you will gain great joy also.

You may not even need to breed replacement chicks and you certainly do not have to hatch off a huge number of eggs. Simply place one or two eggs under her for she will receive the same enjoyment from rearing this small clutch.

If you really do not want any more chickens, you could always hatch some off for a fellow keeper. Ask around your poultry-keeping friends and let them know you have a broody hen ready to sit on a nest. You will be surprised how many will jump at the chance to use your real life incubator.

The one downside I have found while hatching my own chicks is the ratio of hens to cockerels. It is around fifty/fifty which, when hatching a clutch of fifteen eggs, amounts to a lot of cockerels. Cockerels can be a pest around the coop as they wear the chickens out with constant chasing and mating. I cannot justify killing cockerels as many commercial poultry keepers do. But I do not want them chasing my chickens around. Cockerels are also very noisy which does not bother me at all as I live in an isolated country place. If you live in a town or close to neighbours, you may get complaints. The early morning call of the cockerel can alert a fox to the presence of your chicken coop. You will also find that cockerels can be very aggressive towards one another. When cockerels fight, the injuries can be extreme and very distressing for the animal itself and the owner.

If you do accidentally end up with too many cockerels and you do not wish to kill them, there are alternatives. Advertise your cockerels in the local press and pet shops and by word of mouth among fellow keepers. Many towns or cities have livestock markets where you will be able to take your cockerels to in order to sell them. These markets usually take place once or twice a month. Again, your local paper should list agricultural auctioneers in the farming sections. Farming and poultry magazines may also be useful for this type of information.

Where do Eggs Come From?

The egg starts its life in the ovary. This is where the yolk of the egg begins its journey with the first light of day (Fig. 19). When the egg reaches the required size, it will be expelled from the ovary. From the ovary it makes its way to the oviduct; it is during time in the oviduct that the two *chalazae* are formed. The chalazae are often termed balancers. They appear like two thin, white cords and they help to keep the yolk central in the egg. As the egg passes through the oviduct, the remainder of the egg is formed. This consists of a membrane that surrounds the yolk and the formation of the egg white. The longest part of the 25/26-hour journey is the entry into the shell gland; this takes up to 20 hours. Once the shell is formed, the egg enters the vagina to receive a coating of the bloom.

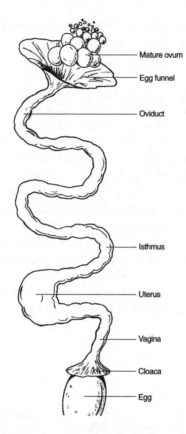

- Mature ovum
- Egg funnel
- Oviduct
- Isthmus
- Uterus
- Vagina
- Cloaca
- Egg

Fig. 19. Reproductive organs of the female chicken

The *vent* of the chicken is used to expel waste and it expels the egg as well. Fortunately the intestinal area that serves the vent is closed off when the egg is about to be passed through. This prevents the egg from contacting any contaminated waste from the intestine. When the egg pops out, you will notice it is spotlessly clean as if it has just emerged from the dishwasher. If your nest boxes are kept clean, there should be little chance of any contamination. If you do see an egg laid, you might just see the shiny bloom before it quickly dries.

How to Spot the Broody Hen

A broody hen is easy to identify, as she sits in the nesting box all day long (Fig. 20). She gathers as many eggs as she can, no matter who lays them. When you look into the nest box, the broody hen will look much larger than her normal size. She will be flatter as she spreads her body out widthways. Her feathers will be ruffled and fluffed out, thus also giving her a much bigger appearance. When you attempt to collect the eggs, she will make some very intimidating noises to try to scare you away. Should you proceed, you are likely to be severely pecked. When you remove the eggs from underneath her, if you are brave enough, she will attempt to roll them gently back under herself to keep them warm.

Fig. 20. A broody hen

She ruffles and fluffs out her feathers.

If you do not wish to hatch off any chicks, then you need to break up the broody cycle. You can remove the chicken to a wire-bottomed cage, well away from her chosen area; the wire bottom allows air to circulate underneath her body. By cooling her underside you hopefully remove her urge for egg sitting. Her food should be reduced, though making sure she has plenty of water. Make sure there are no nesting materials in the cage as this could help prolong her broodiness.

If your chickens are true free-range, you may well be in for a surprise. Chickens often collect a clutch of eggs in a secret nest that you may know nothing about. The chicken just appears one day with her brood, and then it is an emergency trip to the feed merchant for some chick crumbs.

As hens lay an egg every day or every other day, it takes roughly ten to fifteen days to lay a full clutch. Bear in mind that other hens may be helping to fill the nest. In order that all chicks hatch at roughly the same time, the hen does not sit until she is sure she has a full clutch. Once she is sitting on her nest, she will only get off for water, food and possibly to defecate.

When the hen is ready to sit, she will do so for twenty/twenty-two days. On the twenty-first day the chicks will start to pip at the eggshell. They will firstly pip through the inner membrane. They will then pip through the shell itself. They develop in the egg with a special little egg tooth which is made up of a horn-type substance that is strong; it grows on the top beak. This egg tooth gives a little more strength to the chick for pipping out. The egg tooth will fall off within a day or two of hatching.

Once the first chick hatches, it can take up to forty-eight hours for the whole clutch to hatch. The hen will usually know when all are hatched; any that she leaves behind will more than likely have been infertile. A word of warning is vital at this point. If you do ever find eggs that have not hatched, or eggs that have been thrown from the nest, do not open them. I have sometimes interfered at this stage and upset myself with the outcome. If the hen comes off the nest with her chicks, discard un-hatched eggs once they are cold. The hen would not have left a live egg, so do not be tempted to open it. Simply place it in the bin and forget about it. If you open it, the egg could simply blow up all over you when the gas accumulated inside explodes. Worse still, you could find a dead chick.

If a hen throws out an egg, try and place it under her again. Eggs that the mother discards are very often infertile but, occasionally, they are fertile. I did once investigate a discarded, cooled egg, which I assumed was infertile, only to watch a little chick die in my hand. So always try to get her to accept the egg back. But if she does not, then there is nothing you can do but throw the egg away. She will have a good reason for discarding it. You could always do an emergency dash to the nearest available incubator, if, like me, you need to be sure you have done all you can. I find that friends with chickens are usually only too willing to help out with space in their incubators.

Breeding Replacement Chicks

Depending upon the breed of hen you have, it is wise to run a cockerel of the same breed with them. This way you will have purebred offspring that you know will supply you with the quality and quantity of eggs you require. Otherwise you could end up with some mixed breed hens that are only small. These mixed breed hens lay delightful little eggs but they consist mainly of yolk. This might suit people who don't like egg white.

With a small flock, one cockerel will be sufficient. The cock mates several times daily, so the more cocks you have, the more exhausting the life of the hens. You will be able to assess the suitability of the cock to hen ratio by the condition of your chickens. If they are showing lots of bald patches on the head and back, you will know there is too much mating going on. Either get some more hens or find other homes for a couple of the cockerels. I have occasionally seen a cockerel pair up with a chicken faithfully. This is very nice to see when it does happen; the cockerel may pick a mate and consider himself paired for life. I have even seen one of my cockerels take on a single mum and her baby with the utmost devotion.

Spring is the best time to think about rearing chicks. Not only is springtime the mating and breeding season for most animals, it is also a time when the weather and temperature are suitable to raise young animals safely.

Once you have purchased your cockerel, introduce him into your flock at dusk. I find this a suitable time due to the quiet and restful atmosphere. Observe initially as you would with the introduction of any animal into an established group or flock. You should be able to tell within ten minutes if there are going to be any problems with personality clash.

If all is quiet, leave well alone until morning, visiting the coop early to see that all is peaceful. Do not assume your eggs are fertile straightaway as it can take a few days at least to be sure. If a cockerel mates with a chicken, then her eggs are fertile within twenty-four hours. However, if you are using eggs from a large flock you will be better waiting a week or two. This way you can be sure that the majority of the eggs are fertile.

Once the cockerel mates with a hen, his excess sperm can be stored in a special internal nest for up to three weeks. This reservoir is located in the oviduct of the hen. If the cockerel mates with the hen and later meets with an unfortunate end, the eggs she lays in his absence will still be fertile. She should lay fertile eggs for three more weeks.

When you are happy that your eggs are fertile (if you have seen lots of mating), you can then begin to collect the eggs you wish to hatch. It is important to collect only good quality eggs for this procedure. Do not go for very large eggs, as there is a risk of double yolks. Though I have not experienced it myself, they are apparently difficult to hatch. A double yolk is technically a twin. If the two chicks develop facing inwards, they may not be able to free themselves from the egg. The chicks would die inside the shell. There would also be the possibility of deformity.

Thin shells may not withstand the hen sitting and moving around on them. Such shells would also be at risk from bacteria entering and thus harming the developing chick.

Don't be greedy and expect your hen to sit on too many eggs. You would need to consider her size and her ability to cover all the eggs completely. Should you put too many under a hen, you may find that as she turns them she may not be able to keep them covered. This would lead to a drop in temperature of the outside eggs, resulting in different development times or even death.

A large breed is capable of sitting ten/twelve eggs. A small breed will only manage half of that. I did once have a hen disappear and return twenty-one days later with a full fifteen chicks hatched off! But as a practice this is not recommended.

You don't have to put the hen's own eggs under her. You can put any fertilized eggs you wish. A hen will brood whatever comes out of the shell (within reason, of course). I recently put two duck eggs under a hen and she reared them beautifully. If you do use duck eggs, take into account that they are bigger than hen eggs so put fewer under her. They will also take a week longer to hatch.

There are two approaches to dealing with a broody hen. Whichever approach you decide to take, it is important that you see she eats and drinks regularly. Change her diet over to scratch corn as once she is broody she stops laying eggs, and so no longer needs the high calcium of layers pellets which could be damaging to her. The corn will also reduce the amount of faeces that she will deposit in the nest. If you are sure she is broody and she is not going to give up, then I would suggest the first option; this is as follows:

In-house Hatching

- Remove eggs from under the hen daily until you decide the start date
- Mark the chosen eggs with a cross in case other hens lay in her nest; mark with pencil to avoid toxin penetrating the shell
- Place the chosen eggs under the broody hen (mark the due date on the calendar)
- Make sure she leaves the nest daily for food, water and toilet
- Check daily for unmarked eggs and remove them
- Feed and water her in the nest if she won't get out (I have seen hens go very hungry for fear of leaving the nest)
- Wait twenty-one/twenty-two days, then gently look under her wings.

The eggs are removed daily up to the start date so that you can choose your best quality eggs to place under the hen on the chosen date. You need the start date to be the same for all the eggs to prevent the hen leaving the nest before all hatching has taken place. If you allowed other hens to continue to lay eggs in her nest, which can happen with this chosen method, you find that the hatching dates are far too variable. This could lead to half developed eggs being abandoned. When chicks are two days old they will leave the nest and the hen will follow. Thus, the eggs left behind will perish. This is why you need to mark the chosen eggs with a pencil. When initially collecting fertile eggs, they can be kept for up to a fortnight without losing their fertility. They need to be gently placed in an egg box or tray and rotated through 90° every day, as shown in Fig. 17 on page 76. This prevents the embryo from congealing and attaching to the membrane of the shell.

Around seven days after the start of your incubation you may wish to check upon the development of the chicks. At the seventh day, by candling the eggs, you should know if the chicks are developing or not. Gently collect the eggs for candling (don't take them all at once as this may upset the broody hen). The procedure is the same as for checking for freshness, and shown in Fig. 18 on page 82. Go into a dark room with your eggs and candling lamp. Place the egg round end down onto the lamp. You should see the blood vessels of the developing embryo. The lamp will show a spider-like image, featuring a small black spot as the spider body and several long legs projecting from the body (the dark leg-like projections are the blood vessels). This image indicates that the egg is actually fertile and development is under way. If the egg shows a transparent image with no projections, you know that the egg is infertile and should be discarded.

By the twenty-first/twenty-second day you should find small chicks have replaced the eggs you left under the hen. They will be drying off and enjoying the nurturing warmth of their mother. They do not usually require food until they are twenty-four/forty-eight hours old. When they are hatched, they still have some of the egg's yolk in their stomach. This keeps them nourished for the first day or two. You must, of course, make fresh water available. By day two when all are hatched I usually move them to a safe house, consisting of a rabbit hutch with a run (you can adapt a safe house from any similar housing). A small fold unit would be ideal if you have one. Whatever housing you decide to use, you must be sure there will be no chance of the chicks straying away from the mother. Keep the brooding area of the housing clean and well bedded. Do not leave food and water in the actual brooding area (this is where the chicks and hen will sleep). Feeding should be done in the run to keep the inside area clean and dry. It is amazing how much chicken poop can be produced during this period.

You need a good depth of bedding material to provide drainage. Chopped straw is ideal or even shredded paper (shredded paper may consolidate so it must be changed more frequently). I would not use shavings at this stage. The chicks may ingest the shavings causing impaction. A combination of sheets of newspaper with a topping of chopped straw works rather well. When soiled, you can simply roll up the newspaper with its contents and replace. The paper also provides good insulation for the chicks.

The above method of brooding means that the hen is not disturbed and upset by being moved from the hen house before the hatch. The only disadvantage is that other hens will lay their eggs in her nest. This should not pose a problem if you remove them daily (remove the unmarked eggs only). When removing the unmarked eggs, do handle the hen and eggs very carefully. Treat her with gentleness and respect as she may find this intrusion stressful. You must also be careful not to manhandle the eggs unnecessarily. Very gently look for the unmarked eggs and remove them. I find that the hens are more settled with in-house brooding as I call it. After the chicks have hatched, you can then move them, and the hen, wherever you wish. By this time maternal instinct has kicked in and is not likely to switch off again.

Chicks are so vulnerable that you dare not leave them with the rest of the flock. They can be taken by rats or even drowned in the troughs. Take no chances and be very protective of mother hen and her brood.

Provide plenty of water in a safe container that will not allow the chicks to be submerged. Provide boiled, mashed-up egg during the first few days and gradually mix in chick crumbs. Do not worry when at first the chicks do not eat. They will copy the hen within a couple of days and soon begin to imitate her behaviour. She will teach them very early on to scratch around for insects and worms.

Make sure that the chicks cannot be separated from the hen accidentally. If separation occurs for any length of time, the small chicks can soon die from the cold. They need to be able to go under her wing as soon as they need to, in order to stay well and healthy throughout the rearing period.

At around a week old, the hen and chicks can be put out daily into the run and locked away safely at night. Make sure that their house is taken into a secure building at night. Foxes can soon break into a hutch and take the chicks out. When a hen is brooding her chicks, I bring her and the brood in every night. I like to secure them in the porch of the house. A garage or shed will be fine so long as it is predator proof.

Using a Brooder Coop

The second option involves moving the hen to a brooder coop without disrupting her desire to sit on the eggs. Occasionally the hen may snub all your efforts to re-establish her brooding instinct in another area, but the following procedure usually works. A brooder coop can be any chicken-and-a-half-sized box, tub, cage, etc. If the hen and her eggs fit in safely and securely, then it should be fine. A recommended size is around eighteen inches (45cm) square. The eggs will be safer and not able to roll away. Again I find rabbit hutches an inexpensive way to provide a good environment for brooding. You could use the rabbit's sleeping compartment. This will give a darker, more private environment. Make sure the brooder coop is kept in a safe environment away from predators. I find that the rabbit hutch is ideal for the first few days until you can place the chicks and hen in a run during the day. Chicks grow at a rapid rate and a rabbit hutch (unless very big) would soon become rather cramped. At a week old you could consider either fixing the rabbit hutch to a run or making/ purchasing a small fold unit depending on your funds.

- Remove the broody hen to a nicely bedded brooding box
- Place under her two or three china or rubber eggs (these are available from your chicken supply shop)
- If she is settled and content after twenty-four/forty-eight hours, replace china eggs with the real eggs
- Provide food and water preferably a short walk from the coop to see she exercises. If she refuses to leave the nest box, then put the food in for her
- Mark the calendar.

Once the chicks are hatched, in whichever method you choose of the two, management is still the same. Removal to safe housing is important and you need to protect both hen and chicks.

Handling Chicks

Once you start rearing your own chicks, it is vital that you know how to handle them. When a day or two old, they are very small and delicate and so must be handled with great care. Using the thumb and fingers, gently place them round the sides of the chick in a cupping action. Raise them upwards in this way. Do not squeeze too hard with the thumb and fingers or you could crush the little chick. Place them in the palm of one hand while gently cupping the other hand over the top to prevent the chicks from suddenly jumping out of your hand.

Rearing Chicks

Rearing chicks is a real treat, especially the first time or two. They grow at a tremendous rate and change in appearance daily. When they first appear, they are tiny, fluffy little creatures oozing cuteness.

Unless you are an expert it is impossible to sex the chicks at a young age. I would suggest that you do not worry about it in those first weeks. Eventually you will see signs that indicate sexes. Females usually develop their feathers at a faster rate than the males but are often smaller in build. The comb on the top of the heads will be bigger and grow more rapidly for the cockerel. You will also notice some challenging behaviour between the cocks as they practise for adulthood. Even as young as three to four weeks I have seen cockerel fights. Nothing serious, just learning behaviour really. Some people like to segregate cockerels early but that is a very individual choice. I like to keep them all together as a clutch. They usually remain very close to each other well into adulthood and beyond.

You will find that the mother hen does a brilliant job rearing her chicks. She will take her job very seriously. I worry when there is a forced separation with such close bonds. I have sadly had one or two such separations usually as a result of the death of the chicks. I have noticed that the hen simply returns to the flock and barely takes a backward glance. Her job has been done to the best of her ability and it's time to move on! If only life were so simple. Chickens show little outward sign of grief but we will never be sure how they really feel.

The hen should be removed from the young birds at about eight/ten weeks of age which is when natural separation occurs. At this stage the chicks are becoming independent and do not require the warmth of her body heat so much. When the chicks reach approximately half their full-grown size at around ten weeks of age, I let the hen and chicks out with the flock. She will wander about staying close to the chicks. Eventually she will leave them for longer and longer periods until suddenly you realize she is back with the flock full-time.

Feeding Chicks

Once the chicks are out in a run daily they need to be fed at least three times per day. The hen will be perfectly fine following the same diet as her young. If you are concerned that you cannot keep up with the food demands, I would suggest you make the food available continuously. You can purchase tiny little feeders and drinkers for the chicks. If you are only doing a little bit of breeding for pleasure, then you are fine using heavy-based bowls that, kept clean, will serve the chicks well. I consider the plastic packaging that is used for minced meat, stewing steak, etc., very good for feeding young chicks. Although not heavy-based, it is ideal for the first few weeks of the chick's life. I collect them when I can and use for feeding mashed up egg and chick crumbs. Remember, of course, that water, when served, must not be deep enough to drown the chicks!

During the first week, feed boiled eggs mixed increasingly with chick crumbs. After a week, you can phase out the boiled egg altogether. Chick crumbs need to be fed up to eight/nine weeks of age. At eight/nine weeks of age they need to be slowly changed onto growers mash or pellets. Any change of diet must be done slowly over a week or so. Do this by adding a little bit of the new food at each meal. Gradually increase the new food and decrease the old. Do this until eventually you have changed the diet. This will prevent any digestive upset caused by a sudden change in the diet. Variety is important even at this young age, so remember to add greens for extra vitamins.

When putting the chicks out into the run be sure that it is moved around every couple of days. Moving to fresh ground not only keeps the vegetation fresh but the danger of disease or parasite infestation is reduced. Make sure that the run has a suitable covering over the top. Chicks are vulnerable to flying predators at this stage. Large birds such as hawks, magpies and crows will swoop down and grab a chick if the opportunity arises. Fresh vegetables can be added to the diet to allow variety and amusement for the chicks, but make sure you do not feed them too many. It is important that they still eat mainly chick crumbs or growers rations. The feed you provide at this stage must be high in protein to help the chicks to grow and develop.

By week sixteen/seventeen it is time slowly to introduce layers pellets. Again, change the diet gradually over a week. Start to add a small amount of the pellets into the growers mix. Increase the pellets as you phase out the growers. By now you will know your cockerels from the hens. The hens will be approaching point of lay. If you do have a high ratio of cockerels, it would be wise to find good homes for them. Too many cockerels will upset the hens. You may also find it difficult to segregate the males. This will lead to them eating layers pellets, which are the wrong diet for them due to the high calcium content. It is also very wasteful and uneconomical. A small fold unit or separate run could be used to house the cockerels until a suitable home can be found. They can be fed on corn and fresh vegetables, etc.

Vaccinations

When purchasing your chickens from reputable breeders they will be vaccinated against Marek's disease, Newcastle disease and infectious bronchitis. However, when rearing stock for your own flock the choice is yours as to whether or not you deem it necessary to inoculate your chicks. More about how to do this is on page 122. If maintaining good bio security and cleanliness, you may not feel the need to inoculate. The vet or a reputable chicken supplier will advise you on this matter and administer medicines accordingly.

Chicks: a Summary

My greatest enjoyment has been rearing the replacement chicks. It is rewarding and very interesting to watch the chicks thrive and grow. Rearing chicks does not come without its problems. Keep the chicks safe from danger by being vigilant at all times. If you keep them in a run, etc., the dangers are reduced dramatically. If free-ranging your chicks in a larger area, make sure it is free from potential dangers such as water containers that a chick can drown in. Incubators are beyond the scope of this book but you could hatch eggs off in an incubator if you preferred. However, you need to purchase an incubator and a brooder to rear them in. Your local supply shop will advise but, be warned, it is nowhere near as rewarding as using the broody hen!

Chapter 7:

Health and Disease in the Chicken

Cleanliness

Just as it is vital to know the signs of good health in the chicken, it is even more important to know the signs of ill health in order to be able to respond to a problem fast. The principle is that prevention is better than cure. We must, therefore, make sure that the environment that the chicken lives in is as close to perfect and as close to nature as possible. Cleanliness is next to Godliness! Keep the chicken coop cleaned on a regular basis; along with all feeders and drinkers. Clean out once a week, especially in the summer months when flies are at their worst. Any old droppings or food left festering will attract flies to lay their eggs, leading to maggots.

Old food and dusty bedding will contain fungus and dust spores which can cause disease of the lungs. Make sure that, if you use hay or straw to bed your chickens, it is dust free. If you buy dusty hay or straw that contains mould spores, inform the farmer and ask for a refund. Most farmers don't supply dusty hay or straw on purpose, and will usually happily refund, or swap for another bale. It would indeed be a false economy to buy such hay and straw, even if it were offered cheaply; over time you would pay a very high price!

Mould and dust cause permanent damage to the lungs of any animal. The lungs are a very delicate structure and easily upset. Once the lungs are under threat of invasion from spores, they attempt to remove the spores or keep them out.

This leads to over-production of mucus in the lung. The lungs go into spasm (as in an asthma attack with humans). They constrict in order to prevent entry by the spores and dust, and breathing becomes laboured. You may not even see symptoms in the early stages but the chickens become unthrifty. Such chickens lack condition and are slow to grow and develop. Their feathers are dull and they are unlikely to gain weight.

Aspergillosis is a disease of the lung, apparent from the conditions described above. In human form it is termed as *farmer's lung*, so named because it is a disease obviously that farmers suffer as a result of working in dusty environments and working with dusty hay and the fungal spores which are present in such hay and feed.

Aspergillosis is also referred to as *brooder pneumonia*. Remember that lung damage cannot be reversed, only managed. We need to prevent this disease by good management. Always use good quality feed and bedding materials and make sure the chicken house is well ventilated.

Dust spores severely affect the human lungs as well. Take care of the chicken by reducing dust and mould and you are also taking care of yourself. Once the lung is damaged, there is no turning back. Unlike broken limbs, lung damage does not repair.

You may be able to find a local gardener to remove your chicken droppings and bedding. Having someone to recycle your chicken manure will help you to avoid recontamination of the bugs and worms into your flock.

Remember that flies and rodents will spread disease if left unchecked. Rat control is a serious subject and must be adhered to. Removal of all old stale food is vital to prevent rat invasion. Whichever litter system you decide to use for your chickens, the litter must be kept clean and dry.

Rotate the ground your chickens have access to. As with any grazed land, it can become sick if over used. Parasites build up over a period of time and land needs to be rested.

During freezing weather parasites will not survive; however, in the warmer weather harrowing the ground (turning over the soil and breaking it down) will spread parasite eggs and larvae. Once spread out in the sunshine they will be burned by the sun's rays; thus ridding the land of the burden. Resting the grazing land is the ideal opportunity to re-seed, adding to the feed value for the future.

Stress can cause disease in poultry, as in any animal and indeed in humans. Keep the chickens as stress-free as possible.

The Five Freedoms

Dr John Webster, emeritus professor of animal husbandry at Bristol University, has enunciated *five freedoms* which have gained international recognition as standards for defining good welfare in domestic animals. They are:

The Five Freedoms

- Freedom from pain, injury and disease
- Freedom from thirst, hunger and malnutrition
- Freedom to exhibit normal behaviour
- Freedom from discomfort
- Freedom from fear and distress

These are the minimum requirements that any animal is entitled to during the course of its working life. Follow the five freedoms and the animals in your care will be at their most stress-free. Lack of any of the freedoms causes stress and lowers resistance to illness and disease.

Keep your finger on the pulse and keep in the know with fellow poultry enthusiasts. Nearby keepers may have information about any disease outbreaks locally; you need to be aware so that you can take preventative measures. These will include a suitable footbath for any visitors who may come into your coop. The footbath needs to contain an appropriate disinfectant that visitors have to walk through. Make sure it is a high quality disinfectant such as *Microcide*. Also, should you visit any of your fellow poultry keepers during a disease outbreak, make sure that you wear different clothing from that which you use to tend your own stock. Wash your hands thoroughly with *Hibiscrub* or similar after any visits. During a serious outbreak of disease, avoid unnecessary visits to other poultry holdings.

If there is an outbreak of a serious infectious disease, take all possible precautions.

You could purchase wheel baths if you have a lot of land and many vehicle movements. The baths were used during the foot and mouth outbreak and have remained in use at cattle markets ever since. The baths must be filled with a disinfectant capable of killing all major germs. Make sure that all your visitors have to pass their vehicle through the wheel bath.

Signs of Good Health

A healthy chicken will stand erect and look interested in her surroundings, as in Fig. 21.

She will move around with the rest of the flock and exhibit normal behaviour such as dust bathing, preening and eating. She should bear weight on both legs equally as she walks about the coop or yard. Her feathers are sleek and shiny and lying flat against her skin. She should be breathing at a normal rate, not panting or laboured. Her nostrils and eyes should be free from any discharge.

Signs of Poor Health

If your chicken is unwell she may be hunched up with ruffled feathers. Her wing feathers may be spanned out and

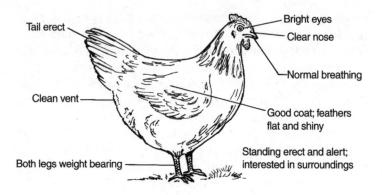

Tail erect

Bright eyes

Clear nose

Normal breathing

Clean vent

Good coat; feathers
flat and shiny

Standing erect and alert;
interested in surroundings

Both legs weight bearing

Fig. 21. Healthy chicken

Standing erect and looking interested in her surroundings.

touching the ground as if she cannot be bothered to carry them. She will appear disinterested and her eyes may be partially or even fully closed. Her tail may become rounded in shape rather than the usual healthy pointed angle. She may have a discharge from her eyes or nose. Diarrhoea may also be evident, causing her tail feathers to be dirty and caked in faeces. She may also stop drinking and eating when feeling unwell. Coughing and/or wheezing may be evident. She may have lost weight.

Should you notice a drop in egg production without any obvious cause, you may have a disease incubating. Watch for any symptoms until egg production returns to normal. If disease does not occur, the chickens may have suffered some shock or disturbance that you were unaware of at the time. Chickens can easily be upset by something that may seem rather mundane to you, such as a severe thunderstorm that could possibly unsettle the birds greatly.

If you suffer a drop in egg production, you should cast your mind back over the three previous days. It takes this length of time for the ovary to stop producing.

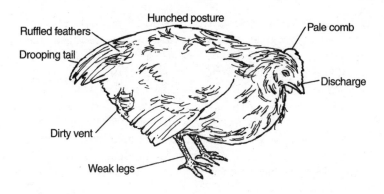

Ruffled feathers
Drooping tail
Hunched posture
Pale comb
Discharge
Dirty vent
Weak legs

Fig. 22. Ill chicken

Hunched up and disinterested.

Should you spot that a chicken is unwell or below par, she must be removed immediately. If a chicken is unwell, she is vulnerable to attack from the strong, healthy chickens. Leaving a sick bird in to die will also encourage the other birds to indulge in cannibalism, which must be prevented at all cost. She may also need to be offered food and water individually.

Remove the sick bird to a quiet space of her own, completely away from the other birds. This will also reduce the spread of anything infectious! Observe carefully the remaining chickens over the following days to ensure there is no contagious disease evident. If in doubt, keep all the stock isolated and bring no further chickens into the group. Keep the sick chicken comfortable.

It could become prohibitively expensive to go to the vet every time a bird becomes ill, but if you have a bit of an epidemic and your vet is good with chickens, it might be worth consulting him or her. I usually just make them comfortable and maybe give them antibiotics or similar. If you find a dead bird in the flock, it could be wise to get it tested to find out the cause of death. The body must be removed as soon as you are aware of it. Leaving a dead chicken in the coop for any length of time can also cause cannibalism, which is difficult to control once it is allowed to start.

Cannibalism: This is a topic which merits consideration. Stress can lead to cannibalism, i.e. over-density in the coop. Make sure you do not keep too many chickens in a small area. Chickens need space to get away from one another should the need arise.

Look out for aggressive behaviour and remove any overly aggressive birds to prevent bullying. This reduces stress on the remaining stock. The bird in question can live in another building or small enclosure (I feel it would be wrong to sentence the bird to death). Even a small hutch would be suitable, provided the bird was let out in a separate run each day. If you have no place to put the bird, maybe a fellow enthusiast would have a suitable enclosure. The nice thing about chicken keepers is they try to help each other out!

Newcastle Disease (NCD): This is highly infectious! It was first identified in Newcastle upon Tyne in 1926. Paramyxovirus 1 causes the disease which is also known as *fowl pest*. A bird could be dead within a day of first showing signs of the illness. Symptoms include extreme difficulty when trying to breathe; they struggle and gasp for each breath they take. The droppings are green in colour and the chicken becomes paralysed throughout its body. Eyes and nose show a discharge.

Fortunately, NCD has been rare in Great Britain for many years and we now have vaccination procedures in place. NCD is a notifiable disease and Defra must be informed as soon as an outbreak is suspected. Sadly, all stock has to be culled and disposed of.

NCD can be spread very easily as it is airborne which means it can pass from bird to bird through the air. Clothes and utensils can also pass on this disease, so great care must be taken if you suspect an outbreak.

Avian (Bird) Flu: Chickens affected by the bird flu virus will usually be dead within forty-eight hours of infection occurring. The symptoms include lethargy and dullness. The chicken loses its appetite and thus condition. The chicken has diarrhoea and also has difficulty breathing.

Droppings and other bodily secretions spread the flu virus from bird to bird. During the migration of wild birds the virus can be carried a great distance. The only way to protect your chickens totally would be to keep them indoors. This is the only way you can be absolutely sure your chickens are not in contact with any infected wild bird. Your chicken house would need to be completely inaccessible to wild migratory birds. A small mesh could be used to cover all holes and crevices that are likely to admit a small wild bird. 25mm wire mesh would cover such holes. Swallows, in particular, are nifty at obtaining entry to even well secured areas. If you allow your chickens access to a run, this would need to have a roof on to prevent wild bird droppings from contaminating the run.

Because of the remote possibility of transmission of bird flu to humans, there have been periods when many bird owners got into a state of panic. I feel very sad that budgies and canaries were being handed in to the local sanctuaries for fear of bird flu. It would be virtually impossible for any house-kept bird to catch the avian flu virus. Chickens were dumped in bags, etc. This is over-reaction, unnecessary and cruel. The local news would certainly inform any area of an outbreak of bird flu. Even then, house birds should simply be kept as per their normal routine with no cause for concern. The only possible concern should be for birds kept outdoors as they are the only type at risk.

Precautions to help prevent avian flu if there is an outbreak:

- Biosecurity is essential; consider footbaths
- Water and feed all chickens indoors to discourage wild birds from feeding
- Use clean, fresh mains water
- If taking in new stock, isolate but better to avoid taking in any new stock at this time
- Be vigilant for symptoms
- Keep all utensils and equipment clean and disinfected
- Discourage wild birds by using scarecrows, flutter tape, etc.

If your chickens are usually free-range, they may take a while to adapt to life indoors should it be required. Keep the chickens clean and mentally stimulated as well as you can. I find that, when forced to stay in (when I know the fox is around), my chickens do seem to adapt within a couple of days. They are used to running out as soon as I open the door, but when I need to keep them in they only try to barge out for a couple of days. After that, they realise that they are not allowed out to play.

Provide stimulation with different types of vegetables and greens. In the summer you can pick lush green grass and dandelions. Be particularly on the look-out for signs of stress. Feather picking and cannibalism are sure indications of stress. Also the chickens may suddenly become bored and eat the eggs, so make regular collections.

Bird flu is a notifiable disease, so the authorities and the vet must be informed immediately!

Coccidiosis is a highly infectious disease. It is common not just in chickens but many other farm and domestic animals, and is caused by microscopic parasites in the intestine. This infection does not pass from one species to another so, if the chickens have it, there should be no risk to other animals. Coccidiosis affects different parts of the intestines and is not limited to one area.

The chickens become weak and lethargic with dramatic weight loss. Feathers will be fluffed up and the chicken stands in a hunched position. Unfortunately, many of the symptoms of chicken ailments are similar; only the vet will be able to give a definite diagnosis. You may, however, notice blood in the droppings. This could well be a reliable indication of the infection. If caught in time and treated accordingly, you should not incur great loss.

If coccidiosis is evident, there are disinfectants available to combat it. *Stalosan F* is a broad-spectrum disinfectant, obtainable in a powdered form. This product is anti-bacterial, anti-viral and anti-fungal too. It destroys, among other parasites, the coccidial cysts that cause coccidiosis. *Bi-OO-cyst* is also available to rid your chicken area of coccidiosis.

Marek's Disease: A highly contagious virus that can spread rapidly through the flock. It attacks the nervous system of the chicken and can cause paralysis. It is the young stock that is susceptible to the virus and one day old is the time to vaccinate against this if you are going to do so. Most deaths from Marek's disease occur at ten/twelve weeks of age. This is why good husbandry is vital. All brooding and rearing equipment must be kept scrupulously clean and disinfected regularly. Particular attention must be applied between hatchings! Once the virus has struck, chances of recovery are virtually nil. The virus can remain for several months in the environment.

Marek's disease is caused by the herpes virus and is also known as *range paralysis*. The nervous system is severely affected and tumours may be present in the internal organs. The virus is passed on from the feather follicles and can be spread easily through the flock. It spreads via the fluff and dust in the shed, from the chickens into the surrounding environment. Chickens do a lot of shaking and ruffling of feathers during the course of a day, thus filling the atmos-phere with dust and skin. The birds inhale the resulting sloughs, thus leading to the possibility of infection. This disease can also travel in the air to neighbouring places without the need for physical contact.

Do not purchase birds that have not been vaccinated against this disease. Any reputable dealer will always carry out the correct vaccinations prior to selling stock. If you do breed replacements, then these too should be considered for vaccination against Marek's (see page 123).

Impacted Crop: This condition is caused by a blockage at the exit of the crop. The exit of the crop allows the food to be passed into the stomach when functioning correctly. If the chicken has been eating unsuitable material, a blockage may occur. A chicken may have eaten some sawdust, feathers or hay. This condition can lead to the death of the bird by starvation. The crop will be enlarged and hard to the touch. Should you find your bird with an impaction, administer-ing warm water to the bird with a syringe could help. Warm water should help to soften and thus disperse the impac-tion. A gentle massage of the area would also help. If warm water does not help, then a dose of vegetable oil or similar may ease the impaction.

If the problem continues with no sign of improvement, the vet can operate to clear the blockage and identify the cause.

The Egg-Bound Chicken: If an egg gets stuck in the chicken, it is vital that you act quickly to alleviate her trauma. Evidence that your chicken is egg-bound is as follows:

- She will sit endlessly in her nest box and produce nothing
- She will be quiet and listless
- Her belly will be distended
- She may smell if infection has set in.

A vet would probably give antibiotics immediately to combat any chance of infection taking hold and try a surgical procedure to remove the egg; this would be a last resort.

Here are some suggested remedies sourced from knowledgeable chicken keepers:

- Wearing rubber gloves, lubricate a finger with mineral oil and gently place inside the vent
- Gently massage the distended area to encourage the egg towards the vent
- Place the chicken's rear end in warm water for several minutes (the water temperature must be the same as the chicken body temperature)
- Keep the bird warm, isolated and comfortable until the egg is passed
- Hold the chicken's rear over warm (not boiling) steam, making sure not to scald her.

I have heard it suggested that you could break the egg inside the chicken. I would not advise this as it could prove dangerous and encourage infection and possible injury to the inside of the chicken.

A similar problem to egg-bind is a *prolapse*. This is usually caused by the passing of over-large eggs or a hen that has been a faithful egg layer for many years. The sagging oviduct protrudes outside the vent, causing the chicken to be vulnerable to infection. You must isolate the bird straightaway to prevent the other chickens pecking at the vent area. The protruding innards must be washed in warm water and gently placed back inside. Some say that if you simply clean and push the organs in every time they protrude, apparently the prolapse should eventually stay in place. A reduction in high-quality food should take her off lay. Place the chicken on a diet of scraps for a week or two. If the prolapse is due to over straining to pass eggs, then withdrawing layers pellets could help.

Scaly Legs: The poultry keeper needs to be vigilant to spot scaly leg early on. A mite that burrows under the scales of the chicken's legs causes this problem. They make tunnels and breed in the ideal conditions created under the scales. During the process, crusty deposits are left under the scales of the leg. These can cause severe deformity of the area. I have seen a Silkie with a totally deformed leg due to scaly leg. The Silkie was very lame and had a massive, badly infested foot. With a good dowsing of Vaseline (petroleum jelly) the crusts gradually peeled off. Sadly the leg was damaged beyond repair and the chicken was left permanently lame. Scaly legs need early detection and treatment to avoid permanent damage. Vaseline or liquid paraffin used to be the main treatment, but nowadays Imervectin is the best drug to use. It's available through vets and farm supply shops, but for chickens has to be diluted to one tenth of the cattle/sheep strength, before use.

Lameness: I have included lameness in this book simply because it is a common problem. I have often found a chicken lame without obvious cause. Bumble foot as mentioned in Chapter 1, page 20, can be one. Perches of a great height are not the only cause of bumble foot; dirty and damp litter can also lead to this condition. The dampness of the litter causes the pads to become tender and sore. The feet are then likely to become infected with the bacteria associated with bumble foot. The foot fills with pus, and then needs to be lanced as soon as possible by the vet. Lancing allows the pus to escape and relieves the pain immediately. Once lanced, the foot needs to be kept clean and free from further infection until fully healed.

Segregate the lame chicken in order to avoid bullying. Chickens are most unsympathetic with ailing companions and can tend to knock them around. A lame chicken may go hungry if left in the coop. Remove to an isolation cage so that not only can her foot be kept clean but also she will be safer until recovery.

Fleas and Mites: Chickens play host to fleas just like many other animals. The flea is usually visible to the naked eye but once detected can jump quickly out of view. They hop from bird to bird with great agility. If your chickens have bald patches and spend a lot of time scratching, you may need to take a closer look. Fleas can actually cause severe weight loss in the chicken if allowed to become a bad infestation. They quickly drain the chicken of blood, causing serious health problems.

With a really bad infestation, the comb and wattles of the chicken may become very pale as the bird becomes more anaemic. Flea powder can be obtained from your suppliers or your veterinarian; this must be applied at the earliest opportunity. Don't forget you can place flea powder in your dust bath or dust bathing area on a regular basis. This will help control the spread of this parasite.

Red mites are tiny mites, red in colour having sucked the chicken's blood. They have eight legs; they are not easy to see as they are so small, however, if you collect your eggs and suddenly feel something crawling all over your flesh, take a closer look. You may see these tiny, red things creeping all over you. They live on the blood of the chicken and usually come out at night. Red mites can soon reduce the condition of your chickens so it is vital to try to eradicate them. They can also carry disease. Try coating the perches, etc., in creosote. Maybe just the underside of the perches and nest boxes would suffice as this is where they hang out in the day. You could also creosote the walls and any other crevices that you think the mite could hide in. Make sure the chicken does not come into contact with the creosote, especially while it's still wet, as this could cause great irritation to her. Too much creosote, even after it has dried, could also irritate the lungs so be careful not to overdo it. Make sure, when using creosote, that ventilation is adequate. If your hen house is concrete you can always whitewash it as this will kill off the bugs.

There are a couple of different and much rarer types of mite such as the *Northern Fowl Mite,* which is a common mite in Australia. The Northern mite lives entirely on its host, similar to the *Black Mite* common in the United States. The *Depluming Mite* can be found in poultry sheds that are not kept clean. Chickens infested with this mite can have bald areas among their feathers. This mite burrows into the skin, causing the feathers to drop out.

A visit to the local poultry supplies shop or the vet should be able to sort you out with a spray or powder to deal with mites. These tiny creatures are persistent and can be difficult to eradicate. You may need to administer two or three treatments before you are able to clear the infestation. If you buy second-hand poultry sheds, do make sure that mites have not inhabited them, as they can lie dormant for up to six months in the woodwork and crevices.

Lice: Chickens find lice infestation very irritating to the skin and this can cause a great deal of stress. The lice can be found on the warmer areas of the chicken as they thrive well in warmth. Louse powder should be obtained without delay and dusted onto the birds and liberally into the nest boxes. Always wear a facemask when dealing with louse and flea powders.

An ideal method to treat the chickens with powders is to add the preparation to their dust bathing areas. If you know where they bathe, then you could sprinkle the powder and leave them to do the work. If you have dust bathing trays or sand-filled trays, use the powders there also.

Worms: You may see evidence of worm infestation in the droppings. If you are supplying good quality food to your chickens, it is a waste of money if you are feeding all those little worms, with the chicken getting very little of the nutrition.

The signs of worm infestation are:

- Diarrhoea
- Poor quality egg-shell
- Loss of weight and condition
- Dull, lethargic birds
- Pale wattles
- Possible death.

Natural Wormers

Cider vinegar is a natural wormer and also a natural antibiotic. It should be administered in the chickens' water daily (25ml to 1 litre of water) to keep them in good condition. The chickens do seem to enjoy the water with this additive.

Carrots are also a natural wormer when purified. The chickens go mad for this remedy. It simply causes them to purge their insides, thus removing the worms. Ground, raw, hulled pumpkin seeds have a coating that will paralyse the worms, so that they can't cling on to the intestine of the chicken. They are then flushed out of the system with ease. Once you have fed the pumpkin seeds, you could then feed either carrot purée or milk. Either of these would cause loose droppings in the chicken, leading to a clear-out of the worms.

Piperazine (the active ingredient of *Biozine*) can be purchased from the chemist or bird supplier, although this is not a licensed wormer for chickens. Mix a 1 oz (28g) serving to each gallon of water. In order to make sure all the birds drink, withhold water for an hour or two, then replace with the medicated water in the chicken house. This should be repeated a week later. It's better to use natural methods, as above, where possible as this medication has a harsh effect on the chicken. After using it, you need to withdraw your eggs from human consumption for a week. Don't throw these eggs away! Have you any chicks that need nourishment? Or maybe another animal will benefit from them. I know that my cats and dogs often love a raw or boiled egg. A boiled egg daily is particularly good if you have a cat or dog, etc., with diarrhoea.

Flubenvet is licensed for chickens and can be obtained from your poultry supplies shop. This usually comes in powdered form and a very small amount (follow directions on the pack) is added to the feed. The fact that this is a powder means that, when added to feed such as layers pellets, the powder will fall to the bottom of the feeder. So you should coat the feed in cod liver oil or cooking oil, the latter working out much cheaper. The coating will cause the powder to stick to the pellets. Give the rations a good mixing in order to coat all the food evenly. The chickens will enjoy having their medication in this manner.

Vaccinations

When purchasing your chickens always make sure that you buy them from a reputable breeder. Go to your local corn merchant and see if you can find a breeder through recommendation. Poultry should be sold with all necessary vaccinations administered. Make sure you check this before you bring the poultry home. The vaccinations required are:

- Newcastle Disease
- Marek's Disease
- Infectious Bronchitis.

The vaccines are available at your chicken supply shop if you do require them. You may have hatched your own chicks in which case you will need to vaccinate them yourself. The vaccines are not overly expensive but there is a lot of waste involved. The minimum dosage I have been able to find is for one thousand birds. So, for a small flock, you would be throwing a lot of medication away. Try to find a local poultry keeper who could possibly help you out with vaccines. If you have poultry-keeping friends, work out a routine that means you can share the product by vaccinating at the same time. Although cheap enough, it is a shame to waste materials.

Vaccines for *Newcastle disease* and *infectious bronchitis* are simple enough to administer; they can also be administered at the same time. They should be given when the chicks are eight weeks old. Keep all chickens in and withdraw the water as the lights go off. They would not drink after dark anyway. Clean the water containers thoroughly and make sure any disinfectants used are well rinsed away. Any disinfectant left after cleaning could destroy the vaccines. Fill the water containers with the required ratio of water to vaccine.

Give the chickens a couple of hours in the morning without water to build up a bit of a thirst. Add the water into your flock and observe for a while to see that they all drink well. Leave the water in and refill medicated water throughout the day as necessary.

Repeat this process at sixteen weeks.

Marek's disease is not so simple to vaccinate against. Marek's requires injections into the breast or the leg muscle (not for the faint hearted). I am sure the vet would happily do the job if you felt unable to inject the animals yourself. Obviously this would add to the expense. If you get someone knowledgeable to help you, after doing the first couple of injections you should be fine.

The vaccine comes freeze-dried, with a diluent that activates it. Once it is activated it must be used immediately and after half an hour it must be discarded safely.

The Marek's vaccine should be given at a day old in order to afford the maximum protection. However, if you have had no previous outbreak of the disease do not worry too much, unless you are buying and selling stock. If you do vaccinate young chicks, then a booster will be required annually to keep the birds covered.

Chickens tend not to be ill for long. They either recover quickly, or they die. Once they have decided they are not well, it is often hard to bring them back from the brink. No matter how well you nurse a chicken, they are vulnerable little birds that give up on life fairly easily. You just have to do your best and try to keep living conditions as favourable as possible. Chickens display very similar symptoms for the majority of diseases. Just keep them adequately fed and watered, in a comfortable well ventilated home, with as much freedom as you can allow. You can't really do any more than that.

Index

A

Age to purchase, 49–50
Aggression, 27, 29, 40, 112
Ark, 36
Aspergillosis, 107
Avian flu, 50, 113–114

B

Back garden chickens, 34–36
Badgers, 29
Baiting station, 70, 71
Bantams, 47–49
Battery hens, 27–28, 44
Beauty treatments using
    eggs, 87–88
Bedding materials, 21–23, 99
Bell drinkers, 13–14
Bird flu, 50, 113–114
Black rat, 69
    Rocks, 46
Boiling eggs, 83
Boredom, 61
Botulism, 60
Bread, 60–61
Breeders, 50
Breeds, chicken, 43 *et seq.*
Bronchitis, infectious, 104,
    122, 123
Brooder, 49
    coop, 100–101

pneumonia, 107
Broody hen, 53–54, 89, 92 *et
    seq.*
Brown rat, 69
Buff Orpingtons, 46
Bumble foot, 20, 118–119
Buying chickens, 50–51

C

Cage, 36
Calcium, 27, 58, 60, 61, 62,
    63, 64, 67–68
Candling, 81, 82, 98
Cannibalism, 29, 40, 112
Carbohydrates, 63
Chalazae, 90
Chicken wire, 25
Chickens, buying, 50–51
    , handling, 52–54
    , quantity of, 50
    , transporting, 51–52
Chicks, 98, 99–100, 101 *et seq.*
    , feeding, 103–104
    , handling, 101
Cider vinegar, 65, 121
Cleaning eggs, 76–77
Cleanliness, 106–108
Clipping wings, 32–34
Cloaca, 56, 58
Coccidiosis, 114–115

Cockerels, 49, 50, 89–90, 94–96, 104
Cod liver oil, 65
Coop, 11, 12
, brooder, 100–101
Cracked eggs, 80
Crop, 55–57
, impacted, 116

D
De-beaking, 27
Deep litter system, 39–40
Dehydration, 66
Disease, 108–109
Dogs, 24, 35–36
Double yolker, 73–74, 96
Drinkers, 13–15
Droppings, 13, 20, 21, 37
Dusk, 32, 35
Dust, 58, 59, 106, 107
boxes, 23–24

E
Egg collection, 73–75
custard, 86
production, maximum, 43
tooth, 93
Egg-bound chicken, 116–117
Eggs, breaking, 19
, chickens eating, 61, 67, 74
, cleaning, 76–77
, cracked, 80
, free-range, 29, 75
, freezing, 77–78
, hatching, 89, 92 *et seq.*
, laying, 17, 74
, pickling, 79–80
, preserving, 78–79
, quantity of, 50

, rotating, 76
, storing, 75 *et seq.*
Eggshell, 61, 64, 65, 67, 80
, baking, 67
colour, 73
Eggy bread recipe, 85
Eglu, 37–39
Electric fencing, 25, 34
Electronic doorkeeper, 32

F
Farmer's lung, 107
Fats, 63
Feather pecking, 29, 40
Feeders, 15–17
Feeding, 15–17
chicks, 103–104
Fencing, 25
Flea treatments, 23, 119
Flubenvet, 122
Fold method, 36–37, 38
Food, 15–17, 55 *et seq.*
, storage of, 59
Footbaths, 109
Fowl pest, 112–113
Foxes, 24–26, 29, 35, 36
Free-range chickens, 29 *et seq.*, 75
eggs, 29
Freezing eggs, 77–78
Fresh eggs, 80–81
Frying eggs, 82, 83

G
Game guard, 65
Gizzard, 55, 56, 57
Grass, 37
Grit, 57, 63, 67–68
Growers mash, 103, 104